The First Re-Animated Division

Carl Hartman

Carl Hartman

My wife, Tammy, and our children, Carloe, Samuel, and Benjamin

To every United States Serviceman and woman who gave their all knowing they could not come back. May they never be forgotten.

Acknowledgment

The Parker Publishing Team. Without them, this would not be possible.

Table of Contents

Chapter 1
The World Is An Unstable Volcano

The world is in total chaos. The War on Terror continues to be waged and now is being fought in every country on the planet. Soldiers on all sides are tired and under-supplied. The last United States President made cuts in the military so deep that American troops are outnumbered nine to one. Better training and equipment are all that have kept the terrorists from winning the war. Soldiers from all sides have become weary of fighting, hunger, homesickness, and burying their comrades. They have not become disloyal; just disconnected. They entered war to save the cause of their country and continue to fight to save their life.

Most small countries are now considered Muslim nations. The American people watched in disbelief as all the allies turned on America. American troops were quickly overrun by the armies of Pakistan, Saudi Arabia, Kuwait, and Iraq. The idea that most Muslims are moderate and should be embraced and welcomed proved to be false. What was, just a few years ago called extreme, is now the actions of an entire religion. America is down to just three allies; England, Canada, and Israel are all that support the United States. Israel cannot provide troops because of the need for soldiers to defend its borders. England once again was decimated by war. The Prime Minister declared that all English troops be recalled to England to help drive out the French and Italians. Canada still had troops in the field, but the public was demanding more and more each day that Canada give up its sovereignty and join the People's Social Equality Union or the PSEU.

The United Nations has been disbanded, and Communism has once again moved to the top of the list as a threat to America. In addition to fighting the Muslims on multiple fronts, The People's Social Equality Union has made progress in Europe. Russia has returned to nearly the same levels of power experienced during the Cold War. Italy, France, and Germany have fallen and every state that separated from Russia has now been reclaimed by the hammer and sickle. Now there is a three-way war. Three different ideas and belief systems are being put to the test and the question of who is right will be answered with bloodshed and body counts. World War III is now at its peak and the very term World War has been redefined. There is not a nation left on earth where fighting is not taking place. Technology plays an important role in fighting, but major battles have come down to fixing bayonets and charging the enemy as well. Hand-to-hand combat is an art of war long ago abandoned by most countries. It is in these battles that the United States has been able to keep up the fighting and claim victories. Battalions have met on the field for this kind of battle and in every instance involving American troops; the Americans have taken the ground.

There was another group of people spread throughout the world. Those people who did not believe in any of the causes. They saw the war on terror as a religious war, and they did not believe in any God. The communists did not believe in any God either. They just seemed to only want to fight. If you were not a communist, then you could only be the enemy of Communism. Individual people did not matter as much as the collective of people.

This group of people suffered the worst from the war. Their villages and towns were in ruins. They were starving and dying from diseases that were long ago thought to have been eradicated. Homes, for the few who had one, were shacks without windows and doors. Those who had electric power were even impoverished according to American standards. Perhaps worst of all was that they were completely rejected by all societies of the world. They were thought of as useless because they would not stand for anything; they would not take a side.

Fighting in this war was a new experience. Many times, soldiers from all three groups met at once. A company of allied soldiers might have the Muslims on the right flank and communists on the left. Both fight with the ferocity of a tiger defending a meal. Once the allied troops were defeated, the fighting would continue at the same level of intensity until only one army could claim victory. All the while, noncombatants suffered and died. Parents hid with their children behind rocks or broken walls. Sometimes parents would hide their children behind themselves. The children's eyes would seem to swell in size from fear with each gunshot or grenade. Their ears were in tremendous pain from the noise level. The sound of aircraft caused them to tremble with unexplainable worry. They have all seen friends die as pieces of flying metal found them instead of a soldier. Parents trying to protect their children would often wonder why. Death by a soldier's gun was almost always quick however, death from starvation or dehydration could take days or even weeks. Still, they would hold their children tight and be ready to offer their own lives if it would save the children during the fighting.

With the wars having gone on for years and the death and destruction growing all around, the noncombatants have taken a side in their hearts. They hate soldiers. All soldiers. The only thing missing is leadership. Someone who can unite those who normally do not get along. Someone who can use wisdom and power to unite the noncombatants with one of the three armies. The Army that best suits their needs.

The United States was in a terrible, previously unknown condition. Though still the 50 United States on paper, only 45 were occupied by free Americans. The PSEU launched an all-out assault on the Eastern seaboard. When the fighting finally leveled off, New York, Maryland, and Connecticut had fallen.

Other Eastern States were decimated and the PSEU held positions in all the coastal regions except South Carolina. With the military spread around the world; it turned out to be private citizens with their private guns who stopped the invasion. Hunters, recreational shooters, peace officers, and criminals became instant soldiers. The number of those that gave their lives will never be known but their contribution to the United States will never be forgotten. New York, Maryland, and Connecticut were close to being retaken as well. The PSEU was struggling with poor leadership and the failing support of the people.

There was a search going on in the Nations of the PSEU for a new leader, someone who could gain the trust of the people and keep them in line, a leader with the ability to talk on the people's level and negotiate with other world leaders.

On the West Coast, California has become a Muslim State. Terrible laws based solely on political correctness allowed it to be captured without a shot being fired. Now if a Californian does not escape, he or she will be forced to declare Allah as God, be killed, or be enslaved. The world's most efficient terrorist training camp is now located in the San Francisco suburbs. Ironically, there are those politicians who see no problem with what has recently happened. Representatives from California argue that things are not as bad as they seem and that the will of the new citizens is to live a life of religious freedom. Representatives from both California and New York insist that soldiers from the PSEU be taken into custody by law enforcement officers and tried in court just the way a citizen would be. There is a call to track down every criminal who fired a shot at the visiting army and put them in jail where they belong.

Perhaps the greatest loss so far is Washington State. Russia, India, Pakistan, and Iran launched every nuclear weapon they possessed. The Ronald Reagan Weapons Platform destroyed all but three in space.

The three that made it through were fired from mosques in California. They were not very big, but they left most of Washington uninhabitable. Millions died from the initial blasts and the radiation that followed. Of all that has happened, nothing has been more shocking than what happened in Washington.

United States soldiers from Washington now realize they do not have a home to go back to. They have fought with a tenacity never seen in combat.

The world is like a volcano. The lava boils and moves just under the surface. Soon, if no relief point is found, it will all blow, and humans will vanish.

Chapter 2
The New Enemy Leader

All humans desire a level of power. For most, having a family or maybe being a supervisor at work fills the need for control. For others, the desire for power becomes life's ambition. The longing to be in control becomes a need. An addiction and the only way to reach euphoria is to have another human ask for permission. Knowing that the success and happiness of a person or group of people can only be determined with the words from the addict's lips is power. Unquestionable power. That is the euphoria Kasmidod Lancegoroff sought.

Kasmidod was born into genuine, abject poverty. He never received medical treatment during his birth. His father was killed in a suicide bombing while trying to get back to the village to see him. Kasmidod barely survived childhood. He was always hungry. He had never left a meal feeling full. Never did his mother ever wonder what to do with the leftovers. He had never tasted clean water as a child. There was only sand-colored water that was cloudy and sometimes difficult to see through. It was all he ever had as a child, but it still always left a bitter taste in his mouth. Sometimes it would deliver a disease to his stomach as well. The memories of walking to play with a friend too frequently ended with him arriving at the friend's shack, only to find that he or she died during the night. Playing caused him and his friends to forget their hunger and thirst for a while. But on the days of the death of a playmate, there was hunger, thirst, and sadness.

One day, five years ago, when Kasmidod was about 12 years old, the water delivered a disease to his mother's stomach. She lay down on a dirty but beautifully colored blanket and never got up. Kasmidod stood there staring at her, shaking her, begging her to get up, but she would not. He noticed how the zig-zag lines in the blanket were different yet somehow needed each other. The bright purple and the orange needed red and the blue to make a complete blanket. The stains on the blanket were just as important. Coffee, the dirty water, blood from a cut on his leg, and his mother's tears; all were needed to make the blanket what it had become. Now his mother's corpse was part of the blanket.

Kasmidod felt like he should cry, but his eyes would not let him. He did not understand death, but he knew it was a part of life. He pulled the edges of the blanket up and laid them over his mother, being careful to be as neat as possible. He looked around for something to take with him, but his heart told him he had all that was needed in his memory. His life's events were woven into who he is now and what made him complete. He knew he needed it all or he would be a different person. And yet somehow, he knew he had many more colors to add.

Kasmidod left his mother and the blanket lying in his shack. He knew that, in a few days, someone would smell her and take her out to bury her. She and the blanket are now the same, forever.

He soon found himself walking on a road. He did not know what road he was on or in which direction he was heading. Soon hunger forced him to his senses. He was sad that his mother was gone but he was over it. The death of so many friends made him that way. For now, he was alive and very hungry. He noticed a fast-approaching cloud of dust coming towards him on the road. A few seconds later, he could hear the whining of an engine. His heart raced. A motor car, he thought. He had seen one once many years ago. At least it seemed like many years ago.

Despite the poverty and the war-torn conditions of the world, some people still had enough power to have a car. This one was getting closer. It was big and bright red. The dust cloud covered all but the front end of the car. The dust cloud could not keep up with the car. It was too powerful and fast. The dust had to submit. Kasmidod's excitement very quickly turned into fear. He remembered that people who have cars are powerful. They control other people, and they are never hungry. They are also brutal. They come and get people, and the family never sees or hears from them again. Kasmidod knew he could be in trouble. He looked for a place to hide but all around him was flat ground. No trees or rocks. Not even a building to run behind. He almost immediately accepted his fate. He was wondering how he was going to die when he noticed the car was slowing down. He imagined how being run over by such a beautiful car would feel when, instead of running over him, it stopped beside him. Kasmidod let out his breath and it surprised him. He did not realize he was holding it. He could not help but smile. The corners of his mouth seemingly touched his ears, and he felt warm and content. He did not know why but he knew his life was about to change. For a fleeting moment, he was not hungry.

As the dust settled down, a man sprang from the car, exiting the back seat as the door was still opening. He ran past Kasmidod and made it about twenty-five feet before he stopped and quickly walked back. Kasmidod stared in such a way that the man knew he was asking what was happening.

"A cramp in my leg," said the man.

He grimaced, grabbed his lower left leg, and walked away a second time. Kasmidod felt comfortable enough to walk closer to the car. It was covered in dust, thin and sporadic in the front and thickening as it moved towards the rear. The red paint could scarcely be seen on the trunk. The car was big and powerful. Kasmidod felt it could run right over his shack and never even get a dent. The engine was still running, and he could feel the ground tremble beneath his feet. The driver's window rolled down and a deep, scolding voice yelled, "We gotta go".

The man was 100 yards from the car and still limping. He did not even respond.

Kasmidod felt compelled by something inside him to clean the car. He took off his raggedy shirt and started knocking the dust from the hood of the car. He could see that the driver was watching his every move with a gun in his hand, perched loosely over the steering wheel. Kasmidod progressed to the driver's door when a button in the shirt Kasmidod was wearing made the sound of metal being scratched. The driver threw open the door, pushing Kasmidod to the ground. He rolled over once and found himself staring up at a very angry man with a gun.

Enough time elapsed for the other man to have walked out his cramp and walked back to the car. Just as he approached the car, the unmistakable sound of a handgun discharging rang out. Kasmidod felt an instant burning pain in his arm. He wanted to cry out, but fear made the action impossible. He grabbed his arm with the other hand and lay as still as he could. His eyes were closed as tight as he could close them, and his teeth were so tightly cinched that the pain in his mouth was greater than the pain in his arm. He asked himself if anyone was going to wrap him in a blanket. Ten seconds later, Kasmidod was wondering why he never heard the next shot. He wondered why he was still alive. Then he could hear arguing. It was muffled but arguing. The words slowly came into audio focus, and he heard:

"Boss, the vagrant scratched your car. You pay me to protect it don't you?"

"You didn't ask him why. It looks to me like he was trying to clean it. I didn't see you cleaning it; maybe the vagrant is worth more than you."

"Come on. You know better. What good is a sickly, ugly boy? Want me to finish him off now?"

"What! You missed at five feet?"

As the argument continued, Kasmidod looked at his wound. There was very little blood. "Just a scratch really," he thought. He left more of his blood on the blanket at home. He raised himself to his feet, the fear having somehow been exercised from his body. He picked up his shirt, carefully tucked the buttons under a layer of fabric, and began wiping away dust again.

"I was just cleaning it, Sir," he said without looking at the men. "Such a beautiful car, it would be even more beautiful if someone would just clean it a little."

The man Kasmidod identified as the boss walked to him and grabbed him by the arm.

"What's your name?" he said

"Kasmidod."

"Why do you want to clean my car?"

"Looks like it's worth cleaning, Sir."

"Name's Karmish. Does your arm hurt?"

"No, Sir."

"You're tough. How old are you?"

"Not sure, Sir. About 11, Mama didn't mark it down and she is dead now."

"Where's your dad? He dead too?"

"Yes, Sir. Killed by religious people."

Karmish turned towards the driver.

"Lazarie. Get over here. Give me that gun."

Larzarie followed the instructions without blinking.

"Now, hold your arm up."

Karmish handed the gun to Kasmidod and simultaneously whispered in his ear.

His heart began to race. There was a new feeling moving into him and he welcomed it without regard for anything else.

"Say you're sorry," he said to Larzarie as he raised the pistol and pointed it at him.

"Or what vagrant, you gonna kill me?"

The next sound was the now familiar explosion caused by the firing pin in the pistol striking the primer on the cartridge igniting the gunpowder. Kasmidod's ears rang in pain and his hand and wrist hurt from the recoil, but he was smiling. The new feeling had now consumed him, and his only thought was "I feel powerful." Larzarie was screaming in pain. Blood was slowly dripping from his upper arm. Not a bad wound but worse than Kasmidod's. Kasmidod turned back to Karmish and handed him the gun.

"I didn't miss, Sir," he said with newfound authority.

"No, you didn't, boy," said Karmish.

Larzarie did not speak a word. He walked to the trunk and opened it. He took out a first aid kit and started to clean his wound.

"The boy first," Karmish shouted.

Larzarie walked over to the boy and cleaned and bandaged his arm. Kasmidod noticed how clean the water was as it mingled with blood and fell to the ground.

He then caught just a glimpse of bread and fruit in the trunk. He tried not to show it, but his youth betrayed him. The look of excitement could not be hidden.

"Ya hungry?" said Karmish

"Yes, Sir," he said excitedly.

"Guess we can eat now."

Karmish tore the bread in two and threw half to Kasmidod. The bread was quickly followed by an apple. Kasmidod stared in near disbelief. He had rarely tasted apples and never a whole one to himself.

"What about Larzarie?"

"That's up to you. You're eating his."

They both laughed as Kasmidod tore off half of his bread and threw it to Larzarie.

"I'm eating the apple," he said flatly.

Somehow, he knew that part of obtaining power was first obtaining trust. Larzarie smiled at Kasmidod and quietly ate his bread. Kasmidod knew in his heart that a new color had just been added to the blanket he was becoming, and he liked it.

Kasmidod was nearly overcome with emotion as he allowed all his new experiences to sink in. He loved that he was recognized for cleaning the car even though that was really for his reasons. He just wanted to see the true color of the car. He relished the fact that an adult had to submit to him. The ability to get immediate revenge on Larzarie for mistreating him made him feel invincible. The most significant feeling, however, came when Kasmidod was left to decide if another human being ate or not. Such control and power had, until now, only been witnessed by Kasmidod. He often wondered how that felt and now he knew. He smiled as he basked in the warmth of his newfound power. He knew he was destined for greatness and Karmish knew it too.

Kasmidod left with Karmish that day. Karmish wasn't particularly wealthy, but he was powerful. He tutored Kasmidod as a father would teach a son and soon, he began to love Kasmidod just as a father would love a son.

Several years passed and it became time to add the final color to the blanket, which was Kasmidod. He was watching a television show about life in America. Television was so rare, even for the powerful because they very seldom could pick up a signal and then they could only be turned on when there was enough gas for the generator. This show was about the poor in America and how little they had.

The reporter was interviewing a woman who was complaining because her free cell phone service was turned off. She was sitting in a car in front of her apartment. In the seat with her were two bags of groceries with food extending past the top of the bags and she was insisting she didn't have enough. The worst came when she admitted she didn't have to pay for any of it.

Kasmidod still had never used a phone. Still ate dinners that sometimes didn't fill his stomach. What is wrong with this woman? Are all Americans this way? He was about to reach a new level of anger when there was a loud explosion followed by gunfire. Kasmidod jumped up and pulled his pistol, Larzarie's pistol. He ran towards the door when Karmish came through the blanket that was the door and fell to his knees. His chest was covered in blood and his skin color was so altered he was hard to recognize. He motioned for Kasmidod and as he approached him he said, "I love you, my son." Blood seeped from the corner of his mouth as he handed Kasmidod the keys to the red car then fell to the ground dead.

Kasmidod cried a little then rage took sadness' place in his heart. Not anger but rage, the result of unchecked anger. Kasmidod jumped through the doorway, the blanket torn down and partially covering Karmish. He fired three quick shots, each landing in the head of three different religious fighters. The shooting stopped for a second, and then three religious fighters from the other side appeared, clapping and cheering. Kasmidod quickly dispatched them in a like manner. His face was drawn in such a way that the anger and hatred might be forever chiseled on it. He held the gun at the ready knowing he had one shot left. From the corner of his eye, he saw a soldier creeping towards him. The soldier was unarmed and bleeding from his hand. Kasmidod pointed the pistol at him and the tendon in his trigger finger began to draw tight. He stopped just short of firing and ordered the soldier to bury Karmish.

It took some time because the soldier's injury was serious, but Karmish was wrapped in the blanket that served as a door and buried behind the shack. Kasmidod then shot the soldier in the back of the head. There was no warning or begging. There was just the report of a gun being fired. People in the village who witnessed the battle ran out of their homes jumping, clapping, and shouting "Kasmidod! Kasmidod!"

Kasmidod ordered the people to collect the weapons and ammunition from the seven soldiers and they did so with no hesitation. He then emptied each soldier's pack and gave the contents to the people. There was food in the packs and the people shouted his name even louder as they ate. Kasmidod now knew for sure how to achieve power. The answer seemed simple now. Feed and protect the people.

Larzarie asked Kasmidod for the car keys and said, "Where to boss?"

"To liberate our people," Kasmidod loudly replied.

He jumped in the car and Larzarie started the engine. Four men from the village secured the weapons and jumped on the hood and trunk. Kasmidod's last color was added, and he was complete. Next were the stains.

Kasmidod drove for days, picking up more and more followers as he went. He also gathered information about the three armies. He knew he needed to align with one and move up in that leadership structure. He knew he must be able to convince the non-combatants that this army was the best suited to meet their needs. He hated religion so the Muslim Army would not work. He could never become a leader in America. That left the army that greatly needed a leader. The Communists, he thought. That was his answer. Kasmidod did move up quickly. His ability to lead was more than he expected. He marveled at how he moved from a boy in a shack to a leader of Nations. Now it was time to exact revenge on the religious fighters for what they did to his village and so many villages like it. It was time for them to pay for what they did to Karmish.

Chapter 3
The Problem and the Problem Solver

The United States Government had been successful at hiding it from the public for the past two years but every day there was a sigh of relief that the war was not completely lost the night before. There were just too many fronts.

The military was fighting in every country around the world. In some countries, there were special operations units but in other countries, there were two or three divisions. Every State's National Guard was assigned to active military units, and the reserve units were all activated. There was still a steady supply of volunteers, but it just was not enough. After nearly a year and six months of political battle, the former president's Social Equality First War Machine Second (SEFWMS) initiative was finally overturned by Congress and now there was money to develop and build better weapons. There were, however, not enough men and women in the field to deploy them. There was a time when it looked like America might withdraw from the war. The President was ready to declare victory and end all fighting for U.S. troops. To make America better, he said, we needed to stop funding military advancements and focus on making Americans financially equal, regardless of their contributions. After some failed attempts, he finally negotiated a law that, in short, prevented any increases in the Pentagon's budget until every American had housing, a stable income, and access to all of life's luxuries whether they earned them or not.

There was still a problem. The shortage of troops. But now there was the money to try to solve the problem. Technology for weaponry was not an issue. Private companies continued their research and now all that was left was to sell it to the government and produce the weapons. The question being kicked around was how to keep troops in the field longer. How could we heal their wounds more quickly and how could we heal serious wounds that sent a soldier home for good? Doctor Kevin Smitherson was selected to find the answer. He was well known for his progressing work on cell regeneration. He led a research team that successfully "grew" an arm and leg for a millionaire who lost his leg in a shark attack. The limbs were successfully attached but the man never really gained full control over them. The project eventually ran out of money and progress grinded to an unfortunate halt. Now there was money. It was military money, and it came with the wholehearted support of the government and the people. Dr. Smitherson was a thin man whose sense of fashion would lead a scientific outsider to believe he could not possibly know what he was doing. A typical day's dress might include red and white pinstripe trousers, a purple shirt, and a light blue sports jacket. He only wore one pair of shoes. Grey generic athletic shoes with Velcro fasteners. He never explained his attire, and he never thought there was anything wrong with it. He needed glasses for reading but wore them all the time. He would just tilt his head down a little and look over the top when he was not reading. He combed his hair every morning but that was all. It was always disheveled at the end of the day. His superiors would schedule meetings with V.I.P.s in the morning so they would only have to make excuses for the wardrobe.

Doctor Smitherson was considered macabre by many of his colleagues. He had no friends, not because he was not likable, but because he did not have time. He was easy to get along with and values and used the advice of his peers but all he thought about was research. He had no family or pets to keep him sane.

When he was not working, he was reading a journal, and he fell asleep each night wondering how he could make more progress tomorrow. Doctor Smitherson was in a deep depression when his funding dried up. Now he was like a rookie race car driver who just found a major sponsor. He did not need to worry about money; all he had to do was drive hard and fast. It did not take long for a person entering Dr. Smitherson's office to know why he was sometimes thought to be odd. In the corner of his office, protected from light by black cloth, sat a glass case. All sides of the case were covered except for the front. Inside the case, and held in place by a silver-colored wire, was what looked like a model of an animal's brain stem. The back of the case was a mirror which gave visitors a view of the whole display. The corner of the room holding the display was noticeably darker than the rest of the room.

The light fixtures above it had been removed. The open space in the case had a purplish glow to it as there was a black light being cast from an unseen source. Sections of the case's front panel possessed magnifying qualities. They were like windows within a window. If a visitor would look very closely, he or she would see strands coming from the model. Strands that were thinner than human hair. If the visitor continued to watch for just a few seconds, a small, barely visible light would appear at the end of the strands. The light would then fade away as quickly as it appeared. The base of the display was a timepiece of some sort. Its numbers were digital but above each section of numbers had white lettering, cluing the visitor in on the meaning of the numbers. Years, Months, Weeks, Days, Hours, and Minutes were the markings, and each had a digital number below them. There was always a short conversation about the display, and it almost always went the same way.

Visitor: "What is this"?

Dr. Smitherson: "What does it look like?"

Visitor: "A model of a brain stem with neurons firing at times."

Dr. Smitherson: "Well, there you go. Very good job."

Visitor: "So why the clock?"

Dr. Smitherson: "What do you think the clock could do?"

Visitor: "I don't know. Measure the life of the battery maybe?"

Dr. Smitherson: "Brilliant, I tell you absolutely brilliant."

The visitor would then smile as if he or she had just won a prize at the fair.

The truth was the brain stem was not a model. Nor was it the brain stem of an animal. It was the brainstem of his brother, Barny Smitherson. A scientist as well, whose work was very similar to Dr. Smitherson's. His last request before succumbing to cancer was for his brother to put their theories to the ultimate test. They believe human beings do indeed have a spirit. There was no formulated opinion of where the soul went after death, but it was there.

Once death occurs, the body is no different than a machine. It needs more maintenance and is much more complicated but if cared for and given a power supply, it could be kept working indefinitely. Ethics is not an issue because the soul has gone wherever it is that souls go. Working with or experimenting on a body is no more unethical than frying the ground-up remains of a dead cow and experimenting with condiments to make it taste better. This peculiar way of thinking made it possible for Dr. Smitherson to experiment with some theories he held on reanimation and regeneration with his brother's corpse. His brother, of course, agreed, before his death.

Dr. Smitherson did reanimate his brother but there was an unintended consequence. His brother tried almost immediately to kill him. It even appeared as though he wanted to eat him. Fortunately, Dr. Smitherson had not yet loosened all the straps on his brother's bed and he was able to secure him.

He found that he could not kill his previously dead brother so, since the soul was gone anyway, he started an autopsy. His brother continued to move and try to get away until Dr. Smitherson removed the brain. Then, on a hunch and as carefully as possible, he separated the brain stem. After several tests, he concluded that the brain stem was still alive. The numbers on the clock below the brain stem in the office showed how long his brother had been alive as just a brain stem and without any assistance. Dr. Smitherson's theory was not very complicated. He believed that the brain stem was responsible for more of human life than modern science gives it credit for. The rest of the brain gathers and stores information, but the stem is what controls everything. The stem can be kept alive when the rest of the brain is considered dead. Information stored in the other lobes of the brain can be moved to the stem in much the same way as a human body grows new arteries from vital organs to the heart when the original artery becomes obstructed. It does not happen in every person, but it does happen. Human beings are programmed to survive. Another important part of Dr. Smitherson's theory was the underestimated power of a neuron.

While it is a known fact that neurons send and receive information, Dr. Smitherson believed neurons also carry the will to live. Unlike other cells that can reproduce, a neuron is forever gone when it dies. Any information the neuron contains dies with it. The human body dies when neurons can no longer transfer information, including the will to live. To most people, a person who died from a gunshot wound to the heart, died because the heart was damaged and could not pump blood to vital organs. To Dr. Smitherson, the person died because the neurons no longer had an incoming fuel supply and therefore could no longer carry the message of survival. Find a way to fuel the neurons in a deceased person and the body could be re-animated. It was not the person who died, just the machine that was left.

He knew he was right. His brother proved he was right. What still needed research was how to control and use the machine and how to get the most information transferred to the brain stem.

Doctor Smitherson's solution to the problem sent to him by the Government was surprisingly simple. His first action was to request volunteers for his experiments. The qualifications were only that the volunteer be active-duty military and that the service member volunteer and their family agreed as well. The family was told that they would not be receiving the remains of their loved ones. The service member was told they would be used to find ways to keep soldiers in the field longer, thus serving their country, even after death. The response was unanticipated. Sixty-five percent of all active-duty service members volunteered. Dr. Smitherson thought he would get around five percent. Service members' families were given a bonus when the service member died, Dr. Smitherson insisted. He viewed it as being the same as a company buying a machine from a supplier. The agreed-upon amount was 5,000 dollars. Service members received a full military burial with all the bells and whistles; there just was not a body.

Research on two fronts was needed. The most pressing was keeping enough soldiers in the field to fight battles. This task would be confronted by Dr. Smitherson personally.

The next was finding a way to safely regenerate tissue so soldiers, and eventually, civilians could heal quicker and better. In his address to his team, Dr. Smitherson even talked about growing new arms and legs. He was also excited that he did not have to work in total secrecy. His results would be on display for the world. The people were expecting results and expecting them soon. His work was classified and his methods were kept in a locked file cabinet but there would be no way to hide the results. Dr. Smitherson never needed attention. He could work alone without ever being acknowledged.

There was something though that he liked about the spotlight. He even relished the stress of deadlines. The idea that the longer his research took, the more people could die was like jet fuel for him. Doctor Smitherson's team was surprised to learn that he had already developed a serum to reanimate a human body. He named the serum Ichor. Ichor was the liquid that flowed through the veins of the Greek gods. He did not fancy himself to be a god or even God-like. He did know there was something special about what he created. Doctor Smitherson made it clear to his team that he would not share the formula with any of them right now. That kind of trust would take a long time to develop. There were just two humans on earth who knew how Ichor was made. One was Dr. Smitherson and the other no longer talked. Ichor was designed around two basic principles of the human body. The first is that the body operates using electricity. The muscles, organs, and bones need calories for fuel but without electricity, everything stops. This is a break away from the belief that calories and oxygen are more important. Dr. Smitherson believed that the need for both could be bypassed, and electricity itself could keep a body functioning. This far-reaching way of thinking had cost Dr. Smitherson many accolades and important positions in the scientific community, but he would not waiver from it.

The second, and even less accepted, was that human beings do have a soul. He would not be drawn into a debate about whom or what made the soul or what happened to the soul after death, but his observations of humans convinced him that there was a soul.

The soul is what prevents most parents from killing and eating their children in the face of hunger. This is just the opposite for nearly every other carnivorous species on the planet. Even the family dog has killed and eaten its young. Most people would conclude that it is not the soul at all that governs that kind of decision but rather the conscience.

Dr. Smitherson argued that many serial killers had been executed while having no remorse and a clear conscience. There was still so much to learn and so many exceptions, but it was his hypothesis, and he was still writing the questions and a means to test them. Of course, one test had already been completed but he could not reveal the results just yet. He thought of how his brother was clearly animated. There was movement of the limbs, attempts to escape, and sounds that were more grunts than anything but distinctly human. His eyes gave the appearance that he knew his brother, but he ignored the family connection and saw him only as a threat and maybe even food. He was reminded of a documentary he once saw of two lionesses chasing an antelope. The antelope inadvertently impaled one of the lions and managed to escape. The other lion, starving and with young back at the pride, dragged the wounded lion back. After only a few minutes of investigation, hunger took over and the pride feasted on its own. The scene faded to black as a male lion was licking a femur. She could have been his sister, a mate, his daughter, or his mother. Now, she was just food.

Dr. Smitherson knew that he would have to overcome this to be completely successful. He would have to know what feelings and emotions were left without the soul. He thought about the traits of a good soldier and how they could be used to make his plan work. Loyalty and valor are two traits he was sure would be useful if they were still there. Could these be used to overcome the drive of survival and hunger? What about patriotism and camaraderie? Were these actions of the soul or learned qualities resulting from years on a team or in a military unit? What of emotions like fear and anger or love and hate?

Dr. Smitherson knew these questions needed to be answered and that the only course of action to get the answers was to conduct the first field experiment. He now had 130 volunteers in the lab. All were deceased service members, and all had the blessing of the families. Money had been exchanged and now he had 130 production machines waiting for him to put them to use. Enough Ichor had been produced, and the military team and equipment were ready. It was go time and Dr. Smitherson was excited. One hundred of the volunteers were dressed in outdated Battle Dress Uniforms complete with boots, a helmet, and a flak jacket. A parachute was strapped to each corpse with only one strap secured under the arms and fastened with a buckle bearing the words "Explosive Hazard." Snickers and jeers filled the air as the task was carried out. The talk was that the bodies were going to be contaminated and used as a biological weapon. Another theory was that they would be used to draw enemy fire as they floated to earth while the live soldiers moved in on foot. The bodies were loaded on Army two-and-a-half-ton trucks and moved to an airfield. On the runway, sat a C-130 with the crew waiting outside the aircraft. The ramp of the aircraft was down and those familiar with the plane could quickly tell that the ship had been modified. There were no tie-down rings for cargo, no seating for crew or passengers and the deck was as smooth as the ice on a frozen pond. Five trucks lined up behind the plane and stopped. Everyone waited for 10 or 15 minutes, and then a car could be seen approaching with several other vehicles right behind it. The car stopped parallel with the first truck and Dr. Smitherson stepped out. He was followed by his team of six scientists and some military brass from the other vehicles. The scientists thought they were going to try a compound that increased the rate at which skin could be grown over burns. They were all dumbfounded when it was realized that none of that material came with them. The crew of the plane was ordered into the cockpit and military personnel security clearances were checked. Once the security check was complete, General Lancer gave the order to proceed.

Dr. Smitherson secured a canister from his car and walked to the rear of the plane.

He then used a remote control and, after a series of codes from the remote, a small workstation appeared from the deck of the plane. He attached the canister to a black hose at the workstation, observed some gauges for a few seconds, and then gave the thumbs up. Next, an Air Force Major approached the station and entered some information using the keyboard. The sound of hydraulic pumps could be heard as small sections of the deck dropped and slid out of sight. The gaps were filled with small metal boxes. Protruding from the boxes were various hoses, straps, and an I.V. kit. Jason counted the boxes. There were a hundred. His mind began to formulate all sorts of crazy ideas, and he wondered just what he had gotten into this time.

While the activity in the back was taking place, the crew was receiving their charts and instructions via a secure link to the Pentagon. They were instructed to fly to California. Intelligence found a new training facility near Oakland. They would be escorted by fighters and assured safe passage and a safe drop zone. Their job was to drop a company of paratroopers into the camp and return to base. Nothing else was told to them. They all knew the troopers they saw on the trucks were unarmed and dead. The subject was fuel for conversation that would last to the drop zone. In the back of the aircraft, soldiers and scientists alike were finishing loading the dead on the plane. Each one was laid next to a box and a strap secured to a tie down on the next box over. An IV was inserted into each dead soldier's arm but it was not secured with any tape. As the last one was being prepared, someone shouted, "They look uncomfortable." Everyone paused for a moment and carefully looked at what was before them. A hundred soldiers are lying in formation, in death, their faces discolored by decay and their bodies bent over the parachutes on their backs. Different body fluids were now collecting on the deck and pooling around the deceased and the smell of decay filled the air.

Dr. Smitherson could sense that guilt and regret were taking hold, so he nearly shouted,

"All right, then. Let's get on with it." He stepped back to the workstation and entered a code. He stepped away and General Lancer stepped up and entered a different code. An audible countdown started with 30 seconds as the security forces charged their rifles and pointed them into the plane. The command was given for the crew to start the engines as the countdown marked 10 seconds, nine, eight, seven, six, five, liquid started flowing through the tubes on the IVs, four, three, two one. "Ladies and gentlemen, I give you the first members of the First Reanimated Division," said General Lancer.

The bodies began to kick and throw their arms around wildly. Groans could be heard over the engines of the plane. Eyes began to open, and pupils adjusted to the light. Helmets slammed against the floor creating the audio illusion of a stampede. A few of the faces looked frightened but most looked angry. Eyes were narrow and stared at anyone they made eye contact with until the living person broke the contact. Lips were drawn back exposing teeth. Dr. Smitherson remembered that look from his brother. His team was supposed to be taking notes, but very little writing was taking place. A video feed was sending images to the Pentagon. Those watching were equally amazed and terrified - only a major event can bring those two adjectives together with such balance and accuracy. This was one of those events. The ramp on the aircraft began to close and the people hurried into their trucks and cars and then drove away. The crew received one last message just before takeoff.

The message was 'Do not, under any circumstances, enter the cargo area,' and they would receive special drop instructions when they neared the drop zone.

CHAPTER 4
A New Bioweapon

The flight to the drop zone was intense. The crew made speculation after speculation about what they were going to drop on the enemy. Not a single member of the crew knew what happened in the back just before takeoff. It was a long ride from Dayton, OH to Oakland, CA, even in an airplane and the crew's imagination had been at work for so long that now fear was starting to set in. Some members of the crew called family members to give them what might be a last goodbye. There was one comfort on the trip. If they were going to die, it would not be from being shot out of the sky. Six F-16s were escorting them and were never out of the range of a friendly AA Battery. They were spaced out with overlapping coverage the entire trip.

Back in the cargo hold, helmets were still banging on the floor, and legs and arms still trashed about as if they had a never-ending supply of energy. However, some of Dr. Smitherson's machines stopped the pointless attempts to get up and began inspecting their uniforms. A few were holding one arm over their eyes and tracing the camouflage patterns with the index finger on the other hand. The cargo area was lit so that better video footage could be recorded. The footage would reveal some essential information. One reanimated soldier, near the center of the formation, somehow realized that he was wearing a parachute and was on a plane. He manually checked the tightness of the strap around his chest, and then began feeling for the straps that should have gone between his legs and around his hips. He pulled up the I.V. tubing, which had long since fallen out of his arm, and stared at it for a few seconds.

His eyes then shifted from one side of the plane to the other as he held up the tubing. After a few seconds, he lowered his arm and smiled.

He shook his head in disbelief and appeared to say some words. These two minutes and twelve seconds of video would soon redirect the research of Dr. Smitherson.

Special drop instructions began to come over the communications equipment as the plane neared the drop zone. It was 200 miles out that enemy jets could be seen. Most were assorted MIGs, but one F-16 was probably captured when California was abandoned.

The pilots of the cargo plane watched with tremendous pride at how quickly the enemy jets were turned into scrap metal. It was almost as if it was rehearsed. The pilot turned to the co-pilot and said, "Guess it is the metal in the man and not the metal in the plane," as the enemy F-16 spiraled to the earth with smoke and fire trailing it all the way down.

The drop instructions seemed very odd but not too difficult. The crew was to fly in at 1000 feet and begin a steep climb just before the drop zone. The oddest part of the instructions was to open the ramp during the steep climb. Neither the pilot nor the co-pilot had ever done this before. As they dropped in altitude, they could see a large military convoy. There were M-

113s and M-577s in a column with Bradleys and Abrams at point and on the flanks. Several deuce-and-a-half and five-ton trucks were formed up in the rear. It was becoming clearer that this was not a "little" experiment. Black smoke could be seen rising from the vehicle exhaust ports, indicating that the vehicles were running and ready to go. The onboard radar of the C-130 showed that there were multiple friendly aircraft in the area.

Apache gunships circled the area, destroying positions outside the camp but nothing was firing at targets on the inside. Resistance quickly dwindled down to just small arms fire coming from inside the camp. It was now apparent to the crew that targets inside the camp were being reserved for whatever was in the back of their plane.

The whining of hydraulic motors could be heard coming from the belly of the plane. The captain radioed for an explanation and was told to ignore the sounds. He shook his head in disbelief, knowing that, as the captain, he should know everything going on in his plane.

What was happening was that the workstation that came from the floor was retracting back into the floor. As soon as that was complete, the straps across the chests of the reanimated soldiers were released and retracted into boxes on the opposite sides. The IV tubes retracted with force, tearing the flesh of the arms they were still attached to.

The boxes then disappeared into the deck, and the undead paratroopers groaned loudly as they tried to stand. Standing, however, was impossible because of the plane's rate of climb.

Three barriers, two feet high, emerged from the floor. The first ran parallel with the ramp and about six inches before it. The other two started at the sides of the plane and extended to the corresponding ends of the first barrier. The formation of the barriers resembled a large V with a flat bottom instead of a pointed one. Some of the corpses started sliding into the barrier as the plane's angle increased. As the plane neared the enemy compound, the pilot opened the loading ramp. The incline and the barriers forced the reanimated troops towards the ramp, and they were "dumped" over the drop zone when the center barrier dropped back into the floor.

Parachutes began to open after two seconds for the pack to spread apart. Some tangled despite the meager effort to prevent it. The ramp of the plane was already closing as the last parachute opened. The deck was now decorated with streaks of body fluids. Reds and grays run down the ramp, getting closer together as they stretch towards their end. At the ramp, there was a pool of thick black sludge. The pool was liquefied enough that it was seeping through the seam created by the ramp and the plane. Small portions fell to the earth while friction pushed out uneven stripes the length of the outside skin of the ramp. The pilot observed several different types of military aircraft circling the compound.

Most traded weapons for cameras. The crew could see the parachutes deploying, but were still oblivious to the fact that the soldiers were now living again. The fall to the earth was short compared to most military drops. The ground fire was heavy but was only provided by individual weapons. Hundreds of photographs and videos were taken during the descent. The photographers were not told about the experiment, and, as a result, many shots were missed as the cameraman opened both eyes to ensure he was seeing what he thought he was seeing. Video cameras were also rolling. Frame by frame, they captured images of decaying bodies wildly kicking their legs and swinging their arms. Heads looked up and down as faces distorted in anger or fear. It was difficult to tell which. Eyes were wide open, and teeth clenched with lips drawn back, exposing discolored teeth and reddish fluids flowing from the mouth corners. There was no preparation for impact or parachute landing falls for 99 of 100 drops. Legs slammed into the ground, causing bones to snap and push through the flesh. A small explosion could be heard each time a trooper touched the ground.

The noise was equivalent to the sound of a .410 shotgun when discharged. The explosion opened the buckle on the parachute strap around the chest. This released the chute, and it was carried away by the wind. Those that were tangled in the air were freed when the parachutes were released. At the same time, enemy soldiers fired bullets into the bodies; the impact causing blood and tissue to jump through the air and land on a different body.

The ride down was much different for one trooper. He had the same facial expressions, but his hands gripped the parachute cord over his head. His legs were together and still. He scanned the drop zone as he descended as if he were selecting targets or cover. He reached down to his thighs as if to release something, his hand frantically moving back and forth and finally giving up and returning to the cord. His face showed his disappointment and frustration as it was clearly visible through the shades of gray that now made up his complexion.

His feet came together, and his toes extended downward just before impact. His toes made contact with the ground, and he reacted to a loud pop. His heels came down as he began to twist his body leftward. The left knee hit with violent precision and the right knee slammed into it as his body now began shifting to the right. His buttocks quickly followed, hitting with such force that it caused him to blink. His legs elevated and he spun around on his back, reaching up with his right hand to release his chute. Again, he was disappointed to find there was no release, but he was relieved to find that the chute was somehow gone. He paused for just a second or two to identify the pain of a possible injury. When he confirmed he felt no pain, he leapt to his feet and began assessing his situation.

His comrades were still huddled together in a mass with a few stragglers still limping or crawling towards the bloody mass. Limbs are now torn from bodies as the energy from a bullet transfers to flesh. The enemy is clearly winning physically but the psychological battle is quickly lost. The mass will not stop moving. The bullets are causing damage that should have killed them all at the onset of battle but none of them were dead. Now they were all grunting and moaning so loudly that it drowned out the sound of gunfire. One by one, they began to look up at the living enemy soldiers. Their faces are horribly disfigured, but determination shows in each one. The look is terrifying to the enemy as it begins to sink in that these soldiers are already dead.

The lone trooper has now realized he is in a firefight. He is compelled by memory to take charge and kill the enemy. He looks for a rifle and punches the wall of a building when he realizes he does not have one. He sees a group of 30 enemy soldiers firing on the mass and he charges. As he runs, he waves his arm over his head and shouts, "Let's go troops!"

He ran through the center of the mass, and as he passed, it followed him. He saw the muzzle flash of an AK-47 and heard the crack as the bullet left the barrel. He closed his eyes in anticipation of falling dead as the slug impacted his abdomen. The impact caused him to buckle over and take a step back, but he did not fall. He felt fine, energized and invincible. A smile formed on his lips as he reached the first Muslim fighter. He tore skin and pieces of flesh from his arms and face. Blood spilled from the wounds, soaking his shirt and pants. He was dead in a matter of seconds. The trooper was inclined to let his kill fall to the ground, but the torn flesh looked so good. The smell of blood was overwhelming him. His face distorted again, this time intentionally, as he opened his mouth as far as he could.

He slammed his mouth on the cheek of the enemy soldier, and he shook his head until a piece of meat was torn loose. He chewed a couple of times, swallowed and repeated the process. Cries to Allah went unanswered as the gunfire slowly quieted and was replaced with screams of terror. The entire group was now feeding in much the same manner. The scene was carnage. Blood was standing in many places, and internal organs littered the walkways of the camp. Soon, there were no living enemy soldiers left and not much was left of the dead. The undead paratroopers were now sitting around as if they had just overeaten at Thanksgiving dinner and could not move. Many were playing with various pieces of military equipment, holding them up and looking at them from various angles.

Their faces were covered in blood and flesh was stuck in their teeth. The sight was dark and evil but somehow the faces looked innocent and puzzled as they struggled to understand why the object in their hand was so familiar and yet completely foreign to them.

The military convoy is now moving into the compound and forming a tight circular formation on the parade field. Helicopters are landing outside the fence, and Humvees are being dispatched to pick up the passengers.

Three 577s entered the perimeter vehicles and drop ramps, and a team of soldiers quickly assembled. Lt. Campton oversees a now seemingly overwhelming task. He and his men have been charged with capturing the reanimated troopers. He has 17 men at his disposal, armed with 9mm pistols and ten-foot-long bite poles, normally used by animal control officers to control dangerous dogs while keeping them at bay. The mission was to capture as many of the troops as possible while they were still moving.

If a living soldier's safety is threatened by a reanimated soldier, the reanimated soldier's brainstem should be destroyed using a pistol or bayonet. The plan was relatively simple. One soldier attracted the target while a second moved in with the bite pole and put it over the target's head and tightened the rope. A third soldier kept watch until the bite pole was in place. The watchman and the decoy then bound the hands and feet of the target with zip strips and attached a thick leather strap over the mouth. The strap was pulled so tightly that it forced the target's mouth open and prevented it from being closed. The team trained for this exercise for six weeks using stand-in targets, so the operation went well until the last two targets were sought. One was found lying face down on a walkway, and the lower back of his head was missing. His brainstem was tattered and torn, causing him to die a second time.

The last target was the same trooper who remembered so much from his days among the living. He now clearly remembered how to shoot a rifle, and he was aiming it at the recovery team. The team drew pistols, but they were in a state of disbelief. There was no training scenario for an armed target. The trooper's face showed the resolve of his intent not to be taken by the makeup of blood. His eyes were narrow and focused on Lt. Campton. His cheeks were drawn tight, creating narrow indentations leading from the corners of his mouth and extending to the ears. His mouth was closed but not tightly and his lower lip was slightly extended past his upper lip. Most of his face seemed so serious and so focused but his mouth looked almost relaxed. As the standoff drifted into the five-minute mark, more and more members of the recovery team joined in with pistols at the ready.

The trooper never took his sight off Lt. Campton. He knew that if he did, it would be all over for him.

He saw the decoy trick and how well it worked so he was now backed into a corner of two exterior walls. He knew he could not win but he knew he would not give up. He decided to shoot the lieutenant and at least take one with him and cause some chaos before he died. Just before his index finger squeezed the trigger on the AK-47, a sergeant turned his body just enough that an American flag could be seen on his shoulder. The trooper dropped the rifle, and his face immediately showed confusion. He felt both of his shoulders and realized he had no patches. He held out his arms and began walking towards the sergeant. The sergeant's instincts told him to shoot but a sense of sadness compelled him to be ready but not shoot until the absolute last minute. When the trooper reached the sergeant, he extended his right index finger and rubbed the American flag as if he were verifying that it was real. His face then changed to a look of extreme sadness. The sergeant recognized the facial expression as one he saw on his son when he found out Santa Claus was not *real*.

The trooper cried out in sadness and stumbled around a bit as he tried to grasp what was happening. He soon walked back to the sergeant, turned around, placed his hands behind his back, and opened his mouth. The sergeant applied the zip tie and leather strap without any further incident.

All 100 reanimated soldiers were laid out in formation on the plane. Their eyes were covered in hopes that doing so would somehow alleviate the squirming and moaning but it did not help. The same recovery team that captured them now guarded them and occasionally had to drag one back into formation.

Once all the reanimated soldiers were restrained and accounted for, other people began to exit vehicles. A HAZMAT team collected the remains of the enemy soldiers. They were searched for identification and all but the three bodies in the best condition were photographed and burned.

An estimated 497 enemy soldiers were killed in the battle. The smell of bleach overtook the compound for a while. Every place where blood was found was disinfected with bleach and water. A hasty Tactical Operations Center was set up and the Bradley Fighting Vehicles and M-1 tanks set up a perimeter around the outside of the compound. The operation was so successful that it was decided to hold the compound and declare it the first step in reclaiming California. A combat brigade from the 82nd Airborne was dropped to help hold the compound until more armor could get there. Engineers were dispatched to build a runway. The plan was for helicopters to bring supplies until the runway was built.

While all the military operations were being planned and implemented, the soldiers of the 1st Reanimated Division were being loaded on trucks and taken to an abandoned airfield. The same C-130 that dropped them into the compound was waiting to take them back to Ohio.

This time though, the crew could see that the soldiers were alive. They fought as they were taken from the trucks and strapped down between the little boxes that were there as the soldiers woke up. This time, there were no IVs, just straps that were tightened so tightly that escape from them was impossible. The crew of the plane had now been fully briefed, and they were also charged with getting all the film and video footage to Dr. Smitherson. The troops in the cargo hold stopped thrashing about and moaning after an hour and they lay motionless for the rest of the trip. Dr. Smitherson was told of his machine's success in battle and could not wait to see it. He was as giddy as a father waiting for a son or daughter to return from a tour in a combat zone. He was also aware of the massive amount of work that needed to be done. Studying photographs, reviewing videos, autopsies and improvements lay close ahead. Dr. Smitherson was also asked to prepare for a press conference.

He would soon be introducing his team and the unclassified portion of his work to the world.

To help him in this task, Dr. Smitherson prepared short biographies about each of his team members. Dr. Jason Jones (JJ) is a young scientist who studied cell regeneration and made significant strides coaxing the brain into producing new brain cells. His goal was to someday revive brain cells or find a way to transfer the information stored in them to new, living cells. Then there are Sam and Ben Manhart. They are brothers whose obsession with weapons and warfare took them both through college and into careers as weapons experts. They are both just as comfortable discussing the designs of primitive clubs as they are discussing the guidance system for a missile carrying a nuclear warhead. Dr. Smitherson did not pick Sam and Ben but they were assigned by the Pentagon.

The team also included Madelan Cavert, a well-known plastic surgeon who achieved medical fame when she discovered a way to grow skin tissue and use it to cover large burn wounds. The difference between what she did and a skin graft is that the skin is "grown" from a few cells. There is no longer a requirement for a second injury resulting from skin being removed. Her work also nearly eliminated scarring. She found that cells surrounding the burn site could be programmed to bond with the new skin and the results were just short of seamless. Madelan left her practice when she learned that the government health care system was not going to pay for her procedure on ordinary citizens. It was just too expensive. Only the elite could afford it. The rest of America had to settle for skin grafts and scars.

Michael Chi is an engineer who turned his skills towards the medical field when his father lost both arms in a farming accident. Dr. Chi is making unprecedented progress in prosthetics that blend with human tissue. Some of his work was already in the testing phase, and the medical community has been quite impressed. One patient has three prosthetic fingers that are completely controlled by the patient's mind, just like the fingers on his other hand.

The final member of the team is Daphne Colman. Dr Colman specializes in neuroscience and has been dedicated to research since graduating from medical school. Her aspiration is to find a way to cure traumatic brain injuries. She admittedly does not subscribe to most of Dr Smitherson's theories, but she could not let the opportunity to work on the project slip by.

Dr. Smitherson was never very good at public speaking and is a little worried about how well he will connect with the world. He jotted his notes about his team on 3X5 index cards and put them in the top drawer of his desk. He thought about going out to buy a matching suit, but his phone rang. It was General Lancer letting him know that the plane would be on the ground in about 30 minutes. He called his team and told them to meet him on the runway in 20 minutes. Looking through his window, he could see that the trucks and security were already assembling, and he knew that the next 30 minutes may very well be the longest 30 minutes of his life. His first one hundred would soon be home.

Chapter 5
They Are Not Zombies!

The plane touched down at the Air Force Base on schedule. Trucks and cars fall in behind as they move down the runway, progressively slowing in speed. Once the aircraft stops and the engines are shut down, the crew is transported to a building out of view of what is happening for debriefing. None of them had much to say except they all knew the soldiers were dead when they first went on the plane, and all but one seemed very much alive when they were put back on a second time. The crew was reminded that they could not discuss the mission and that all questions should be deferred to General Lancer and his staff.

The loading ramp was opened remotely. The smell was nauseating, and the people at the back of the plane were quickly overcome by it. The HAZMAT crew was the only people whose gag reflexes did not respond violently. They were all wearing masks in preparation for cleaning the plane. Flies immediately swarmed the plane and began the process of reproduction. Efforts to shoo them away are as productive as efforts to rid New York's subways of rats. You can kill a few, but you can never get them all. The reanimated soldiers are all still held in place by the straps. All of them are lying motionless with their eyes closed. Dr. Smitherson's heart sinks as he panics inside, thinking that they all must have stopped working. His fears, however, are just as quickly alleviated when soldiers start walking up the ramp. They have no masks and are only wearing military issue leather gloves. The odor clearly bothered them, yet they continued their ascent up the ramp. With his shirt collar pulled over his nose and mouth, Sam asked one soldier how he could continue his task.

The soldier replied, "Used to it Sir". Sam unwittingly asked, "How can you be used to it? This was the first mission". The soldier had progressed halfway up the ramp by this time, but he stopped, paused for a moment, then quickly turned and walked down the ramp towards Sam. Security forces drew their weapons and aimed them at the soldier but did not fire. The soldier put his face about three inches from Sam's and said in a completely different voice that was louder and at a quicker cadence, "Carrying my brothers and sisters off the battlefield Sir". He just as quickly turned away and ran to catch up with his team. The soldier's face left an impression on Sam's mind. He looked so young, yet the lines and scar revealed he had been through too much. Sam felt warmth move up his face as embarrassment and regret caused him to blush. He resolved in his mind that he would never again question how a service member goes about their assigned task.

The reanimated soldiers started shaking their heads and groaning as the crew neared the top of the ramp. Eyes opened under the masks as if they could smell the approaching men and women. The team completed their task with very few words. Two soldiers would approach a reanimated soldier, and one would depress a button on the little box. One would secure the feet and the other would place his or her hands under the armpits and carry the load down the ramp. They would hand the wiggling body to a crew in the back of a truck, and that crew would set them on a bench and strap them to the wall. This task went on without a hitch until the last moving body was placed on a truck. All the way in the back, in the left-hand corner of the plane, was one more reanimated soldier. An Army Specialist leaned over the body and pushed a torn piece of flesh back up on his face. The Specialist bowed his head for a second, then waved for his squad leader. Operations were halted as the squad leader sent a soldier to the column of vehicles. He emerged from a HUMVEE with an American flag.

Dr. Smitherson asked what was going on as the soldier ran past him with the flag. The soldier did not stop but instead ignored the doctor and ran up the ramp and to the squad leader. Eight soldiers gathered around the body and began ceremoniously unfolding the flag.

Dr. Smitherson moved to the bottom of the ramp and demanded that the soldiers stop. His reasoning was that this soldier had already received this honor and was now no more than a wrecked jeep. The soldiers stopped for just a second while the squad leader shouted, "We will do this my way, or you can come carry him off, Sir". General Lancer started towards the ramp, but Sam held out his arm and stopped him. Dr. Smitherson looked at Sam for a second, then lowered his head and shouted, "Continue Sergeant!". The soldiers stretched the flag out over the body and tucked in the edges.

They then carefully lifted him as though he were only injured, and too much jarring could worsen his condition. The

Specialist led the procession down the ramp, his face appearing chiseled from a block of granite, except that tears were slowly forming and flowing down his face. Madelan asked the Specialist what was wrong because she did not know anything better. The procession halted and he said with anger in his voice. "I have carried this man off the field dead before. His name is Robert, and he has, or had, two kids and a wife." The procession continued and they placed Robert in the back of a HUMVEE instead of the truck. Dr. Smitherson was angry with himself because he did not foresee something like this happening. The bodies of the three Muslim Fighters were loaded on a separate truck and the convoy disappeared into a cluster of buildings.

The convoy entered an aircraft hangar, and, one by one, the vehicles were taken by a very large elevator ten stories deep into the earth and into an enormous laboratory. The lab included several offices, a holding cell and two large, walk-in refrigerators. There is also a small theater and state-of-the-art audio and video equipment. The lab portion of the room contains some of the most advanced scientific equipment imaginable. Three autopsy tables are the most prominent. The stainless steel reflected light back, making some parts of the tables invisible from certain parts of the room.

The steel is new, having never been contaminated with blood. A dozen or so assistants and workers are inspecting the equipment. They are clad in contamination-proof uniforms that make them look almost inhuman. Crews are already loading video into the video equipment while others are starting the process of getting the reanimated soldiers of the trucks. The soldiers are awake and restless now, and their stench is filling every corner of the facility. It is the unmistakable smell of death that can turn even the strongest stomach if it is caught off guard.

Many of the people in the room are putting menthol salve under their noses to cover the smell as a civilian worker runs to a numerical keypad on the wall. He shouts, "Sorry, I forgot," as he enters some numbers into the keypad. The sounds of several large fans can be heard as a slight breeze begins climbing from the floor to the vented ceiling, taking much of the smell with it.

The soldiers have reached a frenzy point now and there is concern that some may escape their bindings or chew through the leather mouthpieces. Their faces look terrifying and evil. Fat and muscle tissue are drying up on the faces and cheek and jaw bones are now prominent features on all of them. It looks as if death and anger have blended and even the hardest of the military personnel working on moving them is becoming frightened.

The reanimated soldiers are being strapped into long benches near the two freezers. Dr. Smitherson catches the stare of one for just a moment and the encounter startles him. Eye to eye contact is made, and fear instantly begins creeping up his spine. He feels uneasy and even anxious. He knows that the soldier is telling him that if he could, he would come to him and kill him. Dr. Smitherson has no idea how he knows this, but he knows it, nonetheless. As the moment ends, he is more frightened by the fact that he does know. He convinces himself that it is the eyes and moves on to his work.

Just a few steps away, one of the reanimated soldiers begins pulling his arms open as hard as he possibly can. A living soldier screams, "he's loose!" as two hands drop to the floor. The reanimated soldier raises his arms to his face in an effort to remove the leather strap, but he is unsuccessful without hands. A soldier draws his bayonet and runs towards the reanimated soldier. Dr. Smitherson shouts out, "base of the skull. Push up and into the brain stem".

The soldier grabs his target by the hair on the back of the head but as he tries to gain control, skin and hair tear away and the reanimated soldier continues to try to free his mouth. The soldier readdresses the target and pushes him into a chair from behind. As the reanimated soldier bends forward over the chair, the soldier quickly inserts the bayonet into the brain stem and then pulls it back out. The reanimated soldier immediately ceases to struggle and stands to his feet. His arms drop to his side, and he steps forward three very small, difficult steps. His face now shows intense fear. He is terrified of what has just happened to him and what will happen next. He drops to his knees, groans as best he can around the leather, and falls forward.

His head comes to rest between his two hands. The soldier who stopped him walks to him, puts his hand on his shoulder, says "sorry brother," then slides his bayonet into the scabbard and returns to his post as if nothing happened. There is complete silence for a moment, then Jason clears his throat and says, "Okay, let's get him on the autopsy table three". Two other soldiers pick up the corpse and carry it to the first stainless table in the row. They put him on the table with his back down and his arms by his side. A different soldier trails carrying the two hands and places them on the motionless chest. The noise level picks back up as the rest of the reanimated soldiers are secured to the benches and the bodies of the enemy soldiers are placed in the smaller of the two coolers. Also in the cooler are the remaining thirty bodies of the 130 original volunteers. One of the soldiers exits the freezer and loses control of his emotions.

He shouts as loudly as he can, drowning out the other sounds, "This is too much. I can't keep working with these zombies". The comments anger Dr. Smitherson. He runs to the center of the room and begins shouting for everyone to gather around him. One by one, the people gather, murmuring as they walk. A half circle is formed around Dr. Smitherson, who is standing on a chair by this time. His hands rest on his hips, a clipboard hanging from his right hand and his thumb wrapped around a pen, pressing it against it.

His face is still red and his tone in both his voice and facial expressions indicates that he is very angry. He says with a fluctuation in his voice that no one has ever heard before. "They are not zombies, and they are never to be referred to as such by anyone on this team again. They are reanimated soldiers. RAS for short, and that is the only title I will accept. Do I make myself clear?"

The military personnel shouted in unison, "Yes, Sir" while civilians mumbled some inaudible words. The soldier who used the word "zombie" was quickly led away by security and was never seen again. The vehicles all left after the short speech, exiting the same way they entered. A HAZMAT crew cleaned them on the top floor before they left the hangar, and they were parked in a motor pool.

Now that things have slowed down, and the sound level has reached normal decibels, Dr. Smitherson can hear his name being called from several different parts of the room.

"Dr. Smitherson. The video is up".

"Hey Doc, the dead one is ready for autopsy. Do you want to do it"?

"Dr. Smitherson. Are you ready to pick out the ones you're going to keep?"

The words become mumbled as a brief but terrifying panic sets in, as the questions keep coming before Dr. Smitherson shakes his head clear and he regains control. His immediate thought is "I'm going to have to delegate some of this. It's too much". He walks over to the benches of RAS and starts counting, carefully examining each one for damage as he goes. The ones that are judged to be in better condition will be saved for the next combat mission. The others are going to be destroyed and used for experimentation. He plans to save 30 and combine them with the 30 in the cooler and 40 brand new volunteers to see how experience and cold storage will affect the performance of RAS units in combat. He gets to the number 30 and starts to walk away when who would be number 31 somehow gets his attention.

Number 31 is virtually unscathed, sitting quietly in his seat and looking straight ahead. Dr. Smitherson placed a bright red 3X5 index card on the backs of each RAS he wanted to keep, and he is thinking about keeping number 31. He realizes that it would not be the actions of a scientist and starts walking towards the monitors to watch the video.

He shouts to his team to start the autopsy on the soldier killed in the lab and instructs them to start on the one killed in action as well. His instructions included examining the brain stem and measuring the rate of decay. Dr. Smitherson needs to know if reanimated soldiers decay at the same rate as dead soldiers. He calls Sam and Ben to the monitors to watch the video with him. He tells them to start thinking of methods to increase the survivability of the RASs in combat. The video starts with footage of the flight to the combat zone. Most of it falls victim to the fast-forward button as Dr. Smitherson is just scanning the video now. He is stunned to see some of the RAS following the camouflage patterns with their fingers. Not all of them but he can count at least 14. At the same time, Dr. Smitherson is watching a video, a civilian worker is walking behind each row of RAS with a syringe and injecting acid into the brain stem of every RAS without a red card taped on his or her back. Death is very quick. Movement stops before the injection can be completed. It is, however, a slow process because the syringes are not prefilled. Each dose must be drawn from a vial. Back at the video station, Dr. Smitherson, Sam and Ben have walked up to a monitor, and they are touching it with their index fingers.

"I'm not sure what he is doing but he is definitely working with the straps", Ben says.

"He can't be Dr. Smitherson, snaps back".

Sam starts to make a comment when a member of the security team runs towards the group.

"He is checking his rigging, Sirs", the soldier exclaims.

"I told you". Ben says like he won an argument.

The soldier points out that the RAS is looking for straps that should be wrapped around his hips.

He notices the mouth moving as if words are being formed and points it out to the team. "Look there Sirs. He is complaining because the straps are not there".

Dr. Smitherson's face goes flush as he recognizes the face in the video. It is number 31. Dr Smitherson runs towards the RAS as he shouts, "I'll be right back". As he approaches the RAS, he can see the syringe sticking out of the back of the neck belonging to number 31. "Stop, " he shouts, as he runs toward the scene with his right hand held high. "Please tell me you haven't terminated this one yet", he repeats three times with a noticeable shortness in his breath. The needle had just penetrated the skin and was just short of the skull. The worker, startled by what is happening, extracts the syringe but in doing so, releases a small amount of acid into the flesh. Dr. Smitherson runs to the autopsy table and snatches a scalpel out of Dr. Colman's hand and returns to the RAS. Knowing something important must be happening, she follows, repeatedly asking, "What's going on?" Dr. Smitherson carefully cuts away the affected tissue. When finished, he walks to the front of the RAS and puts a hand on his shoulder. "Are you alright?" He asks, with his face inches away from the face of the RAS. The RAS looks puzzled about being asked a question but tries to answer. He cannot speak because of the leather strap, so he nods his head up and down three times. Dr. Colman's mouth gaped open wide, letting the pen she placed in her mouth to assist Dr. Smitherson fall to the floor. "Did he just respond to your question?" she asked, stuttering her way through it. "I think so", Dr. Smitherson replied with an ear-to-ear grin. Each person returned to his or her post and continued with their assigned task. Dr. Smitherson decided he would never assign termination of an RAS again without first watching the video footage. The team worked for 16 straight hours.

The next day, a meeting was called for 3:00 PM so everyone could review their notes and develop a presentation for the team and General Lancer. There was not a lot of time set aside for rest because the next combat mission was already planned. This time, it would be against the communists in New York and the Pentagon is hoping for another victory. Officials are also expecting to see some use of weapons in the next assault. The team was given two weeks to get ready for the next round of fighting. Two weeks to make what was a nearly flawless victory in California even more impressive in New York.

Chapter 6
Analysis, the Reports, and the Fallout

The Analysis

The review of the mission and subsequent experiments, autopsies and developments were revealing. The most startling fact came from the video. The way the RAS attacked and ate the enemy was a surprise to everyone except Dr. Smitherson and even he was taken aback by how powerful the drive to consume flesh was. One RAS was hit by 17 bullets as he approached six enemy soldiers. He began feeding while others were still shooting him. It was as if he never even noticed the bullets impacting him.

Another unexpected finding was the strength the RAS troops possessed. They could rip flesh open by just grasping it and pulling it; when they did, it appeared to be effortless. The RAS soldier that was killed was caught on video as well; the team counted 11 hits as he approached his target. The Muslim soldier threw down his rifle and drew his sidearm just as the RAS grabbed him and bit deep into his neck. He put the pistol against the base of the RAS's skull and pulled the trigger. The RAS fell to the ground, rolled a few times and stopped moving. The wound he inflicted with just one bite, however, caused the Muslim soldier to bleed to death.

The star of all the video footage was, however, number 31. Dr. Smitherson brought in six members of the 82nd Airborne to watch the video and to comment on everything they could about number 31's performance. The footage from the plane was easy for the team to explain. Staff Sergeant O'Connel pointed out that number 31 was checking his rigging. He was making sure all his straps were tight so he would not fall out of his chute when it deployed. He points to the screen and says, "Look, he is angry because there are no leg straps. He is mumbling something; probably complaining about Government Issue and laughing off his anger". A large crowd is now assembled around the monitor as SSG O'Connel continues. "Here, where he is scanning the sides of the plane, he is looking for the static line. This man knows he is about to jump, and he is aggravated because things are not the way they are supposed to be".

"Is he doing things correctly?" Sam asks.

"He is," replied SSG O'Connel. "See how he is holding on with his hands above his head after he checks to ensure he has a canopy".

"Amazing," Dr. Smitherson says in a soft, quiet voice.

SSG O'Connel points to the screen and puts the video in slow motion. "Look right here.

He is trying to drop his ruck, and he is upset that he can't find it".

"Ruck?" Dr Smitherson asks.

"Backpack. He would need to drop it if he had one, so it doesn't interfere with his PLF".

"What's a PLF?" asks JJ.

"Parachute Landing Fall". "It's what keeps legs from getting broken.

Look at the screen. See how his landing looks practiced. He looks stiff but it is correct.

The energy created by the impact is spread from his toes through his heels, knees, and finally his buttocks and released with a partial spin at the end. He wants to release his chute here". SSG O'Connel followed the entire landing with his index finger against the monitor and "pausing" several times as he spoke.

The team watched as 31 jumped up and moved to cover. Not because of fear or cowardness but to survey the battle and plan his attack. The soldiers watching the video pointed out the fact that 31 learned the bullets could not stop him and used that to his advantage. The crowd was silenced, except for audible sighs and gasps when 31 spoke. He clearly was calling for and trying to lead his fellow combatants. Dr. Smitherson was caught entirely off guard when he saw this. He had to get up and leave the room, his face going from serious with his hand on his forehead to shaking his fist in excitement and laughing, then back to serious again. His knees were a little wobbly as he walked around the corner of the partition separating the video equipment from the rest of the room.

"Pause it! Pause it for a minute," he shouted as he left the room.

When he returned and the video continued, it was pointed out how 31 looks for a weapon.

He is upset that he does not have one, but carries on with his assault anyway. The last important bits of information gleaned from all the videos were that even 31 had to eat just like the others and just as savagely. He tore off flesh and ate it before his opponent was deceased. It was difficult for even the battle-tested soldiers to watch.

The neck, face, and head seemed to be favorite food sources for RAS but in general, any chunk of flesh would work. The brain seemed to be the favorite organ. Muslim soldiers not wearing helmets had their skulls broken open and their brains dislodged and eaten. A few RAS troopers discovered that the helmet could be removed, exposing the top of the head. This new skill leads to more enemy soldiers having their brains consumed by RAS troopers.

Dr. Colman pointed out that the RAS would quit eating after the enemy soldier had been dead for 15 to 20 minutes, regardless of how much flesh remained. It was not because the RAS soldier was finished eating; in fact, they would often go kill another enemy soldier. She thought this needed further research.

Then there was the rest at the conclusion of the mission. Seeing toothpicks and cigarettes would not have been out of place if they were living, breathing soldiers. They however, were not. They were clinically dead and had no use for eating or resting. There should be no scientific reason for them to attack and kill other human beings, and yet they do.

Dr. Smitherson ordered his team to answer some questions. Why do they eat? What happens to the food? What motivates them to pursue other humans? Why do they rest when no food is available? Why do they prefer flesh from the head and brain? He felt these questions must be answered.

Dr. Smitherson also ordered the development of a safety system to be used if civilians and friendly troops encounter RAS. Methods of repairing the damaged machines are also an important project.

He needed to know if they could be used more than once, if they became damaged. He gave the team the "what" and left the "how" up to them. His last request was to the Pentagon. He took 31 fingerprints and sent them to the Pentagon with a message stating he would need the complete 201 jacket to continue. He needed to know what it is that makes 31 so different from the rest of the volunteers.

Fall Out

An unexpected video surfaced online as the team pored over the videos, data, and autopsy reports. The terrorist training camp had several security cameras rolling during the raid and the images were watched and recorded at a separate facility. Some of the cameras captured images that the team recording for Dr. Smitherson missed. The images were gory, and they brought to light an unforeseen problem. In one section of footage, a Muslim soldier is pushed up against a camera by two RAS. He is clearly begging for mercy and trying to surrender, with his arms in the air as he stumbles backwards. He is not granted any quarter, though as the two RAS begin biting his face and neck, his screams in both pain and horror drowned out the gunfire that was previously dominating the air. He called out to his comrades, but his cries were unanswered. He soon stopped screaming as parts of his skull and spine became visible. Security cameras record other acts of carnage, but this one quickly becomes both a rallying cry and a political tool.

King Abdullah of Saudi Arabia demands an apology and an immediate halt to the project. He was at least expecting a public apology since former President Obama had so freely given them in the past.

He also called on the countries of the world to condemn the United States for its evil and demonic weapons. "Allah", he said, "would not approve and would utterly destroy America". The condemnation came quickly but was useless since all the condemning countries were already at war with America. President Arubio however, offered no apology. Instead, he issued a warning to both King Abdullah and Supreme Czar (SC) Lancegoroff to surrender or more of the same would follow. American troops have been fighting with ridiculous rules of engagement and no protection or fair treatment when taken prisoner and it was now time the battlefield was leveled. He warned that you should lay down your arms and run if the reanimated soldiers came to your unit. "America will win this war on all fronts," he said with renewed vigor and confidence.

That, of course, did not sit well with what was left of the left. A new antiwar campaign was started and the images from the enemy's security videos became the Democratic talking points. Senator Mikevell from California called for a Supreme Court Review of the repeal of the SEFWMS initiative. His premise was that the initiative was an executive directive and could only be repealed by a member of the executive branch, namely the President. The initiative, he argued, was repealed by a Republican majority Congress, making it illegal. His argument was considered weak, but the Justices decided to hear it. To appease Sen. Mikevell, President Arubio repealed the SEF-WMS but the Supreme Court invalidated his decision, stating it was pending a hearing so that no action could be taken. The court ruled that funding could continue until the matter was decided.

Sen. Mikevell also stirred up a sleeping anti-war movement in the people. Public protest became commonplace again and the movement was growing. This time, the movement had the support of the Catholic Church. The Catholic Church decided that the reanimated soldiers are an abomination to God and the human body itself. The message was that man, and science went too far in trying to give life to a person of whom God has already taken life away. This was an unexpected response from a church whose home was destroyed and a Mosque built in its place, but it was the church's response, nonetheless. Many of the Catholic soldiers who volunteered to become reanimated soldiers withdrew at the request of the church. Other churches found the video of the RAS in action troubling and ghastly, but concluded that the soul was already gone. Reanimating the corpse for continued service was acceptable if the service members volunteered. It was, in a sense, no different than deciding to be buried in the ground, at sea, or be cremated. What is being reanimated is just the shell and nothing more. The greatest and most effective political fallout came from China. China is a member of the PSEU and manufactures most of the weapons used in the war by PSEU soldiers.

President Chowlin petitioned the World Court to freeze all the assets of the United States of America and to seize the assets for China until a debt of 17 trillion dollars is repaid. Chowlin knew this action would not affect most of America's assets, but his plan was to cause enough disruption to force the defunding of reanimation research. He saw the real threat presented when his troops would have to kill American soldiers twice. He understood that it was difficult enough to do it the first time around.

President Chowlin scheduled a meeting with scientists and SC Lancegoroff to discuss the use of any funds recovered from the United States and the development of new weapons that could effectively repel attacks from RAS units. To this point, the only defense being deployed is to hide anti-aircraft guns around military bases and fill the air with flak so the RAS troops can parachute through it. This was a dangerous task because the RAS troops were dropped at such a low altitude that the flak would endanger the PSEU troops almost as much as the RAS troops.

King Abdullah took a completely different approach. He had Muslim clerics declare that Allah would grant any believer killed by a RAS immediate access to Heaven. This rallied the troops and strengthened the hope that future attacks could be overcome. The King also declared that every American soldier killed in combat was to be dismembered, rendering them useless for the reanimation program. His thinking was that if nothing else, it would work psychologically against American troops and it would impact the number of soldiers that must be killed twice. The King took one other important step. He sent most of the Muslim forces from the training facility in San Francisco to retake the camp at Oakland. A victory there would reignite the hearts of the troops and lessen the psychological effects the RASs are having on his troops.

The fallout that affected Dr. Smitherson the most had nothing to do with politics, strategy or religion. Something happened that caused even him to doubt the rightness of his works. Millions of people saw the video released by the Muslims, including the families of the RAS volunteers. Children were being interviewed because they were told their mommy or daddy was dead and yet they saw them walking around on television. They looked sick but alive. One little girl tore at Dr. Smitherson's heart. She was asked by a reporter what she thought about seeing her daddy on TV and her reply was more than Dr. Smitherson was prepared for.

"I was told my daddy was dead, but he is still alive and fighting. He looks so mean now that I'm afraid of him. He can't love me anymore since he never called to tell me he was still alive."

Mothers and fathers, as well as husbands and wives of RAS shared the little girl's view. How their loved one could be dead and still walking on the Earth and fighting in a war is so confusing. The cry for honest answers was growing stronger and stronger. Some families of volunteers began demanding their loved ones' remains.

For the second time since the project began, Dr. Smitherson is angry at himself for not properly preparing and, as a result, hurting others. He did not like the feeling it brought with it, and he was genuinely sorry for the hurt he caused. He knows, however, that if America is going to last, he must continue. He called his team and informed them all that all work stops until the public knows exactly what he is doing. Not the details but what their loved ones are volunteering for. He did not want to be responsible for another child's heartbreak. General Lancer argued briefly but saw the resolve in Dr. Smitherson's eyes and conceded. Dr. Smitherson could include the information in his upcoming press conference. He allowed his team to go back to work.

REPORT 1

General Mark Lancer

General Lancer's report was short. He first said how impressed he was with the decisive victory in Oakland. He knew that the United States now had a formidable weapon that could destroy whole enemy units or send them running in fear. The report includes four recommendations from the Pentagon:

Brass is concerned with how close RAS troops must get to the enemy to be effective and how vulnerable they are to injury. The recommendation is that a weapons system and better armor be incorporated into each RAS troop, to include a method to prevent the legs from breaking when they are dropped in.

Active-duty service members are excited about the victory in Oakland, and even those who disagree with the program are caught up in the debate of which branch of service produces the best RAS troops. To increase morale, each volunteer entering the program must have a chest patch on the right front breast pocket that identifies the volunteer's former branch of service. An effort will be made to count kills and release the data to military personnel, if only for morale's sake.

The RAS needs a Division Patch, or so says the Pentagon. A contest was held among wounded service members in Walter Reed Hospital, and General Lancer displayed a slide on the wall with the winning design. The patch was four inches long and two and a quarter inches wide at its widest point.

It is in the shape of a classic six-sided coffin and is dark grey. A bright red number one, two and a half inches tall is centered in the middle of the coffin, and the words "SECOND TIME" are arched on the top in three-eighths-inch letters with white thread. On the bottom of the coffin is the word "AROUND" in the same fashion but in a reverse arch. The letters resembled an oval with the middle taken out. The number one does not look like the one made famous by the First Infantry Division. It is thicker and does not have a base. The soldier who won the design contest was a member of the First Infantry Division and a future volunteer for the First Reanimated Division.

Finally, the Pentagon wanted the RAS tested in a defensive role. The Pentagon was searching for the ideal conditions to secretly replace undead soldiers with RAS units to see how they could defend a position and hold ground.

General Lancer went on about his belief in the program and how he was so proud to be a part of it. He added that he thought very highly of air dropping the RAS into battle because it all but eliminated enemy artillery. He shared the intelligence reports about the increased use by the enemy of anti-aircraft artillery. He assured the group that a counterplan is in the works and would be ready for the next mission.

REPORT 2

Sam and Ben Manhart

Ben spoke for the two while Sam changed slides or held up visual aids when they were called for. Ben started by acknowledging his agreement with General Lancer's stated need for armament and better armor. An image appeared on the projector. It was an M16. Not an M16A1, A2, or A3 but an M-16. Pictured with it were some cables and two unidentifiable pieces.

There was also a small circular cage-looking device. Jaws dropped as sighs and murmurs filled the room as Ben began to explain the image. He made several requests for silence so he could speak. "This," he said, "Is an M16 salvaged from a weapons storage facility in Kettering.

It is a basic design and is simple to operate. Our plan is to attach these rifles to the less dominant hand and arm of RAS troops. It will be held in place by the manipulating cage. The cage has several servos and motors that will manipulate the rifle so that it fires on enemy targets accurately. The two other devices you see in the image are a thermal scanner and a laser system. In simple terms, it will work like this. The rifles are attached along with the thermal scanner and laser. The same system that releases the parachute will activate the weapon. Once armed, the laser will constantly scan for heat sources using the thermal scanner. Once a source is detected measuring 90 to 100 degrees Fahrenheit, the system will start the servos and motors working to bring the rifle to bear on the target. A small thin wire will be placed in contact with the nerves that control the index finger on the firing hand. A small electric charge will be sent to the nerve, causing it to contract the muscle and fire the weapon. Each RAS will have 30 rounds of ammunition. In testing, the system has a target hit rate of 27 out of 30 targets. Once all ammunition has been expended, the system will detach itself from the RAS trooper and fall to the ground. It can be picked up by cleanup crews or destroyed remotely by commanders to prevent enemy soldiers from capturing a system. The system does have two potentially serious flaws. Collateral damage if there are civilians in the area, and a danger to friendly troops if a friendly unit is nearby. Sam and I will demonstrate the system on the live fire range at 1300 hours tomorrow."

Questions begin to fill the air just as Ben gets out his last words, but he chooses not to answer any of them. Instead, he calls for Sam to hold up an odd-shaped bowl made from metal.

"This is, I believe, the answer to survivability for RAS troops. It is made from titanium and will stop a round from an M2 HB at 100 yards. This is the Replacement Skull Cap (RSC) for RAS troops. It covers the forehead and extends across the top of the head, reaching down the back of the head to cover the upper neck. The brain stem is protected as much as possible while not interfering with what has proven to be the RAS Trooper's greatest weapon, his or her teeth. The RSC will be molded to fit each RAS and permanently attached to the head with surgical screws and wire. A Kevlar helmet will still be worn for greater protection. The system's ability to protect a human head will be demonstrated immediately following the weapon demonstration tomorrow on the live fire range. "

"Finally," he concludes. "There is the issue of broken legs during an airborne insertion." Sam holds up a flat piece of metal bent in two places to make it look like a long, narrow letter 'U.' This device will be placed in the boots of each RAS. This piece of steel will provide support to the lower leg without interfering with the trooper's ability to walk. It will not prevent all breaks, but it will decrease the number of breaks. Questions will be answered at the demonstrations tomorrow to allow time for each team to give a report tonight". Sam turned off the projector, and he and Ben left without another word. Neither one liked public speaking. They would flip a coin to determine who had to talk and who got to demonstrate. When they were finished, they were just that, finished. Staying around and risking a question would just prolong the uncomfortable feelings, so why stay around?

REPORT 3

The Medical Team

Dr. Jones, Dr. Cavert, Dr. Chi, and Dr. Coleman

Dr. Coleman spoke for the team. The rest stayed seated with the audience. She decided to give her report in response to the questions Dr. Smitherson asked the team. Her facial expressions indicated that she was very excited before she uttered her first word.

"Ladies and Gentlemen," she began.

"We believe we have discovered why RAS troops eat human flesh and so much of it."

"Why do they become idol when no food source is present?" "Why do they prefer flesh closest to the brain?"

"And finally, we believe we have discovered how they continue to function with no blood flow and without oxygen being provided to the muscles and organs."

Dr. Coleman explained that when a human dies, all the organs stop working. This had been the accepted belief for hundreds of years. Tests showed, however, that the brain stem continued to function at a very low level even after decomposition starts. All that was required was a supply of energy to increase the level of activity in the brain stem, and "survival genes" take over. The requirement for nutrition was still present, but the source had changed dramatically. The need for calories, vitamins, and other nutrients was replaced by a need for electricity.

The operation of the human body was dependent on electrical impulses sent from the brain to other parts of the body, telling them what to do. Stop the flow of the electric impulses, and life stops. When the human body is alive, there is also the requirement of oxygen and energy created when calories are broken down in the digestive system and supplied to muscles and organs via the blood. Most of this process occurs involuntarily, but still would not occur at all without electrical impulses from the brain.

Once the human body had died, the requirements changed to just electricity. Unlike in the movies, it was not the same as the electricity provided by a bolt of lightning or plugging a cord into a wall. What was needed by the brain stem was the electricity generated by the human body. Since the organs were no longer providing the fuel the body converts to electricity, the brain stem eventually died, and death became irreversible. If, however, a suitable energy supply became available, the brain stem would take over and use the body to seek out more fuel or food. Dr. Smitherson's Ichor Serum, which was still classified, provided the energy needed to boost the brain stem. It took over the operations of the body and reanimated it. When a reanimated soldier consumed an enemy soldier, he or she was not eating for calories or nutrients. The flesh just sat in the stomach and rotted. What happened was that the brainstem of the reanimated soldier sent neurons to the stomach. The neurons extracted human-produced electricity from the flesh and returned it to the brain stem. The brain stem used the energy immediately or stored it for future use. This was why reanimated soldiers seemed to prefer feeding on the brain. While neurons traveled throughout the body, the greatest concentration was in the brain. When a reanimated soldier was biting the face and neck of an enemy, he or she was trying to get to the brain.

The brain stem ran a very efficient operation. The reanimated soldier could walk, fight, use all his or her limbs, see, smell, and swallow; all while using very little energy. In fact, the amount of energy was so minuscule that we had to design a new meter to measure it. To help you understand how small the amount of energy was, imagine one bulb on a miniature Christmas light set.

Paint a picture in your mind of the bulb while it was illuminated. Now, dim the light produced by about 500,000 times. If your painting was correct, you couldn't even tell if the bulb was lit now. That was how much energy a RAS trooper needed to operate during a combat mission. Much more research was needed to determine how the reanimated body functions on such a small amount of energy. This knowledge also explains why the reanimated soldiers will not eat other living organisms or drink liquids. They have no nutritional value to them, and energy would be wasted trying to consume them.

"Perhaps more amazing than the low energy requirement is the way RAS troopers can conserve and store energy. We have discovered that when a RAS appears to be sleeping, he or she is conserving energy. There is no movement or brain activity, and the only sense that appears to be functioning is the sense of smell. In one experiment, RAS were left isolated in a dark room until they ceased movement. After twelve hours, a soldier was sent into the room. The RAS almost immediately started trying to escape their bonds and get to the soldier. The same experiment was repeated, except that a dog was released into the room, and the RAS remained quiet and motionless. It appears chemicals released by humans into the air can be detected by the RAS's olfactory system, and that the presence of those chemicals brings the RAS from a state of rest to a near-immediate alertness. We have not yet discovered what chemical or chemicals the RAS can detect, but a simple experiment was conducted to prove that it involves the olfactory system. The system was completely severed from the brain in two RAS troopers and then removed. Neither could detect a human standing less than three feet away unless the human was standing in front of the RAS in a lighted area or making a very loud noise. Humans could also walk right by the two RAS at less than one foot while they were in a state of rest, and the RAS did not respond. It is known that the neurons in the olfactory system are the only neurons in the body that can reproduce, so it is believed that this influences the advanced use by RAS troopers. The system has adapted for survival."

Dr. Coleman went on to explain that there are most likely other chemicals or a set of circumstances that could reanimate the dead. She pointed out the history of zombies as it relates to science. There was some ancient history that supports living beings being afraid that the dead would rise again. Illustrations of the living dead were just as horrific as how zombies are portrayed in today's movies and video games. The goal of the resurrected dead seemed to be the same. Find the living and eat their flesh.

Dr. Smitherson's face turned bright red at this portion of the report. In part because of anger at the use of the word "zombie," but mostly because of the embarrassment he was feeling. He, just at this very moment, realized he did create zombies. He shook off the feeling with the thought that he had done something more than just create a zombie and continued to listen to Dr. Coleman.

Dr. Coleman continued with comments about the need for continued research on methods of fueling RAS troopers without the requirement of human flesh. She did mention, however, that this research is meeting with strong resistance from the Pentagon. Military officials are concerned that if the need to hunt and kill humans is removed with an alternative fuel source, the RAS will no longer be combat-effective. Dr. Colman admitted this was a validated concern if the only plan for reanimation is to create combat zombies. She made it clear that she was looking to the future and other potential uses for reanimated humans. Security and service in dangerous situations were a couple of examples she threw out to the audience as food for thought.

She concluded with some comments about safety, repairing RAS troopers, and how the team's research was being used in the field to help the living. Each operational RAS had been fitted with two devices attached to the brain stem. These devices did not cause discomfort because they couldn't feel pain. The first was a small explosive charge that could be detonated by field commanders should RAS encounter friendly troops or civilians. The charge was enough to destroy the brain stem, making the RAA non-operational and preventing injuries to others.

The second device was a small charge of compressed carbon dioxide and a coolant. This could be released to cool down the brain stem. We discovered that cooling the brain stem to near freezing immobilizes the RAS. It was during this time that they could more easily be repaired or recovered from combat missions. Broken bones did not heal in RAS, so we developed plates and pins to be used by medical personnel in the field to make repairs. These pins and plates would stabilize broken bones until they could be permanently repaired with bonding agents.

A lab-grown skin was developed to be used on large, open wounds, but it did not work. The living tissue could not bond with the dead tissue of the RAS. We concluded that the treatment for large wounds on an RAS trooper was duct tape. It sealed the wound, and the bond was long-lasting. The tape, or any other medical equipment for that matter, did not need to be sterile as RAS troopers were immune to infection. The artificial skin developed by the medical team had proven to be very effective in the field on living soldiers.

The tissue of the soldier bonds with the skin around a wound, and the unused portion of the artificial skin falls away as part of the healing process. This had been especially helpful for soldiers suffering from large burn areas. The healing time had been cut by nearly a quarter, and scarring was minimal and in some cases, nonexistent.

We also adapted the device designed by the Manhart brothers for use on the inside of prosthetic limbs. Though still in trials, the application looked promising. Several of the soldiers in the trials had been able to hug and lift their children for the first time in months or even years. The adaptation was complicated on paper but simple to explain. The device still detected heat sources, but instead of a laser sight, the user guided the artificial limb towards the target. Sensors allowed the user to move the prostheses at the joints and secure the target. The sensors allowed the brain to send the message to move in much the same way that messages were sent in living tissue.

Dr. Coleman introduced the medical team a second time, and they all exited the platform, and the audience dispersed in different directions, talking in different groups as they left the area.

Dr. Smitherson was pleased with the progress and was looking forward to the next combat trial. He ordered the installation of the RSC on all RAS troops scheduled for the next combat trials. He also made a note to speak to Dr. Coleman about her insistence on mentioning an alternative fuel source for RAS troops. He was not opposed to the idea, but he was concerned about funding for the program if the military thinks its money was being spent on research they did not want to pay for. Dr.

Smitherson then went to his office to work on his press conference speech.

Chapter 7
Number 31

The sun was shining brightly. So brightly, in fact, that an occasional beam of light could be seen reflecting from a blade of grass. There was not a cloud in the sky, and birds sing from the few trees that had not been scorched by the flames of war. A large dog came into view, nose to the ground in a desperate search for food. The dog carried on with its search despite a large burn wound on its back and hindquarters. It found a piece of some discarded bread and pounced on it. The dog placed the bread between its front paws and raised its head, turning it from side to side, searching for competitors. Once the dog deemed it safe, it lowered its head and quickly consumed the bread, licking the blades of grass to ensure nothing was left.

The dog put its nose back to the ground and restarted the quest for food when a loud whistle broke the calm. The dog raised its head and scanned the area, but did not see anyone or anything. It put its head down, but before it took the next step, "Here, boy," "Here, girl," filled the air. The dog seemed to smile as it cautiously trotted towards the sounds. The calls were repeated several times until the dog was standing in front of a well-fortified defensive fighting position with two faces staring at it through the opening in front. The dog was overjoyed as they laughed and petted its head. It lay down and human hands vigorously rubbed its belly. The humans were laughing like schoolchildren. They were just as excited to see the dog as the dog was to see them. One human reached into his rucksack and pulled out a Meal Ready to Eat (MRE). He tore into several smaller packages. He opened each package and dumped the contents on the ground in front of the foxhole. The molecules of ham and eggs lifted into the air were drawn into the dog's snout with the very next breath. Her playful joy changed to survival as she lunged towards the food that was still spilling out of the pack.

The human moved his hand just before it was bitten with eggs and ham. In just a few seconds, all the food was gone. The dog even licked up the freeze-dried coffee and creamer. The human collected the garbage, putting all the smaller packages back into the larger one and putting the larger one in the foxhole. "Female German Shepherd," broke the silence. "Yeah. A big one at that!" was the reply.

"You shouldn't have given it your food, Sarge. It might be two or three days before we get more."

"I don't think we'll be here in two or three days, Johnson."

The conversation was interrupted by someone shouting.

"Sgt. Lowe, Sgt. Lowe. They want you at the TOC now!"

"In route," Sgt. Lowe replied.

Sgt. Lowe climbed out of the fighting position, telling Pvt Johnson he would be back in a few minutes. He began walking towards several M577s with canvas extensions out and people moving about in organized chaos. The scene resembled a campsite, with each camper having a tent extension on the back. The dog took up a position on Sgt. Lowe's right side and followed him to the TOC. Other soldiers asked about the dog, but did not receive an answer. Sgt. Lowe did not have time for conversation. He knew he was about to receive operation orders for his squad, and he took that seriously, too seriously to hold up a briefing for conversation about a stray dog.

A sign welcoming visitors to Lockport, New York, was visible as he walked towards the crowd. Tensions were high as everyone knew the information could not be good. Their unit patches were from the Second Ranger Battalion, Eleventh Armored Cavalry Regiment, and the Tenth Mountain Division.

These units passed through Canada in what was going to be the retaking of New York. Other units moved on New York from the East and South and progressed relatively unchallenged, but Kasmodod moved most of his troops North, hoping to flee to Canada if needed. The resistance at Lockport had been much higher than anticipated by United States commanders. Now, after three weeks and four days of heavy fighting, all that remained were remnants of a Ranger Battalion, a light infantry division, and parts of an armored Cavalry brigade. Less than 4000 troops faced an unknown number of enemy soldiers, and neither side drew up plans for surrender.

Both sides were out of small arms ammunition and food. Supplies could not get through because there was no fuel for the vehicles, and cargo planes from either side were shot down by the other side. The time was near for someone to take the ground, and it was going to be with bayonets and hand-to-hand combat. The 11th ACR had M1s and BFVs but no fuel to move them to the battlefield. The vehicles were abandoned, and the crews got a quick refresher course from the rangers and light infantry in rifle bayonet fighting and hand-to-hand combat. Every soldier knew what was coming, and now squad leaders were getting the orders for the battle.

Captain Jefferies from the 2nd Ranger Battalion presented the operations order. It was short and to the point.

"Men and women, I do not need to tell you the importance of what we are about to do. We are going to attack an unknown number of enemy soldiers with very little or no ammunition and with no incoming supplies. We will force march about six clicks and charge over a hill at the enemy. They know we are here, and they know we are coming. We will have very limited artillery support and zero air support.

We know that the enemy is out of beans and bullets, too, because we have intercepted radio messages and carriers requesting immediate resupply. Enemy strength can only be estimated and includes numbers that are even with us, up to as high as out-numbering us three to one.

So, what do we have? We have three thousand eight hundred and fifty-seven of the finest warriors God ever placed on this planet. Nothing places greater fear in the hearts of our enemies than seeing United States military personnel charging towards them with a bayonet leading the way. Other officers are telling their troops the same thing I am telling you. We can win. We must win.

The plan is simple. All units will form a line roughly two kilometers long and move towards the enemy. Once contact is made, the center will hold while the ends move up flanking the enemy and forming a giant "U". The flanks will then collapse on the enemy, leaving them the only options of fleeing like cowards before the flanks collapse or die. Remember, you are the leaders. Remind your troops that we do not want to die for our country. We want to give the enemy the greatest opportunity to die for theirs. Now, get your soldiers ready to fight. We move out at 1700 hours, and troops, we are the center of the formation. We will prevail!" Several loud and robust Hoorahs could be heard as the squad leaders checked their watches and started walking back to their fighting positions.

Sgt. Lowe arrived at his position with the dog still on his right side. He bent down and extended his hand to Pvt Johnson while at the same time telling him to get out of the hole and gather the squad. Johnson moved from position to position, telling the occupants of each foxhole to rally on Sgt. Lowe. Soon, there were 16 battle-weary and battle-hardened faces staring at him, intently waiting for his instructions and guidance. All of them showed signs of some type of injury. One had three bloody stubs where fingers once were. Not one complained, whined, or cried. They all just listened.

"We are going to assault the commies before they can get supplies and reinforcements. It will be a direct assault, and your weapons will be your body, your mind, and your bayonet. Remember to fight in threes with your backs towards each other whenever possible, so your backside will be protected.

We move out in three hours. Fill in your fox holes, eat what you have, and meet right here at 1645 hours, hoorah?" A loud "hoorah" answered back 18 voices strong. "Who are we?" shouted Sgt. Lowe. "Rangers," came in response. "What are we going to do?" shouted Sgt. Lowe. "Kill them all and let God sort them out." The group dispersed back to their positions and began disassembling and filling in the fighting positions that had been home for more than a week. Some ate while others sharpened their bayonets and fighting knives. Soon, the distinct sound of steel being pulled across a smooth stone filled the air. There was an occasional voice, but the sound of steel on stone was dominant. Confidence filled the air despite what would clearly be desperate times by any other person's standards. Not a single soldier of the 3857 soldiers preparing for battle on the American side planned to die today, and yet they all knew that many would fall.

1645 was now showing on Sgt Lowe's watch, and all his squad were assembled and ready. The squad spread out at double arm intervals so that the man on each end of the squad was standing next to a member of a different group.

Sgt. Lowe looked to his left and right through his binoculars and could not believe what he saw. In either direction, and for about a kilometer, was a row of soldiers. There were men and women, privates and captains, wounded and healthy, but all highly motivated. Most had rifles with a bayonet fixed on the end. Those who carried sidearms had now moved their 9 mm pistols around to the other side of the pistol belt and put their knife in its place. The pistol would still be good for use as a blunt weapon should the knife get lost, but right now, a good, sharp knife with a heavy handle was second best only to the bayonet.

At exactly 1700 hours, every soldier took the first step towards the enemy. Faces were already tense, even though there was still a six-kilometer march ahead; the troops were ready to fight now. The line moved surprisingly straight for being so long, despite the differences in terrain features. As the line approached the fourth kilometer, the dog barked and ran ahead. She stopped about twenty steps in front of the line and sniffed the ground. She then turned back, sat down, and barked at Sgt. Lowe. Sgt Lowe's face went flush as he made a fist with his right hand and held it high over his head. Everyone in line repeated the signal, and the line came to a halt, the ends having taken only three more steps than the center. Captain Jefferies walked up to Sgt Lowe, and quietly asked him why he stopped the line. Sgt Lowes pointed to the dog and told the captain that he believed the dog had found some explosives. Sgt Lowe called two soldiers to him and instructed them to use their bayonets and probe the area marked by the dog for mines. The two soldiers took the bayonets off their rifles and started walking towards the dog. When they got to the spot, both lay down on their stomachs and very carefully started pushing the bayonets into the ground at an angle. A metallic sound was heard, and one soldier shouted, "Got something here."

He withdrew the bayonet and carefully began removing grass and dirt with his hands. Soon, three prongs were exposed, and the soldier recognized the Bouncing Betty mine. He carefully removed the dirt from the top of the mine and saw that the arming pin was left on top. He carefully put the arming pin in place and loudly sighed as his shoulders fell in a moment of relaxation. He cautiously pulled the mine from the earth and carried it over to Capt. Jefferies and Sgt. Lowe.

Sgt. Lowe told the captain that he had an idea. He got a piece of 550 cord from his rucksack and tied a loop in one end. He then pulled the other end through the loop to create a lasso. He walked up to the dog, put the lasso over her head, and pulled it tight around her neck. "Search," he commanded, and the dog carefully but quickly began sniffing the air and ground, stopping and barking every meter or so. Sgt. Lowe marked each place with pieces of MRE packaging, and in 10 minutes, the dog found 16 more mines.

Captain Jefferies sent a message across the formation telling everyone to take up a prone position to lower their profile and for all commanders and squad leaders to meet him in the center of the formation in 30 minutes. He then told Sgt. Lowe that his dog just potentially saved hundreds of lives. Sgt. Lowe took a diary out of his pocket and recorded the incident. He made a note to recommend the dog for a medal when this battle was over. Captain Jefferies was writing as well. The discovery of the mine field changed his plan and his tactics. He sent two runners back to the TOC with handwritten orders to send back all the 100-mile-an-hour tape and cardboard that could be found with the runners. He asked the runners to complete their task as quickly as possible. Thirty minutes later, the captain shared his new plan with his officers and squad leaders.

He explained his belief that the mine field was more than just a defensive weapon. He believed that it was an early warning system and that the enemy did not know how far we had progressed, or maybe even that we had started our assault, and he planned to use that to his advantage. He first ordered the squad leaders to have their people probe for mines to determine how wide and deep the field was. He predicted that it should not be too large because the enemy was low on supplies. He ordered Sgt. Lowe to walk the line with his dog, looking for mines and marking them. He wanted only men and women who felt comfortable disarming and removing the mines to do so. An explosion would give away their position. He then wanted all the mines brought to the center of the formation. He explained that his new plan would delay the attack until the next morning, but his new plan would increase the chances of success. He ordered six, three-man teams with flares sent forward two hundred meters as early warning teams and told everyone to work quickly, quietly, and safely.

Sgt. Lowe and what was now apparently his dog found many more mines. The mine field turned out to be just two rows deep, and though concentrated in the center of the field, the mines were spread much more thinly on the edges. Their placement seemed to confirm Capt. Jefferies assessed that they were there more as a warning device than anything else. A total of 60 mines were uncovered. The arming pins were left with the mines so the enemy could recover them and use them again. Six Claymore mines were also found, complete with trip wires and electrical firing systems. Everything was brought to the center of the line, along with the tape and cardboard brought back by the runners. Captain Jefferies again assembled the leadership and gave them the rest of his new battle plan. He picked up a piece of cardboard and used a knife to cut out what appeared to be crude stabilizing fins from a bomb.

He then taped them to the bottom of one of the mines. He held it up to his eye level and examined it closely, smiling from ear to ear after a few seconds. "All right, troops, here is the new plan. We are going to make 50 of these newly invented, fin-stabilized, hand-thrown mine bombs. I'll rename it after the war. We are still going to use the "U" formation, but we are going to bring them to us and ambush them. It just might be the biggest ambush in military history. Tonight, you will move your people into position with the 60 bombs spread out at even intervals along the "U". Issue the bombs to the strongest throwers in your units. Use of the bomb is simple. Remove the arming pin and throw it up and towards the enemy like a football. They are heavy, so throw hard. The fins will cause the bomb to land on the nose first and detonate the mine. The killing power will be diminished because the mine will shoot the charge into the ground instead of the air, but I believe there will still be enemy casualties. The primary goal, however, is not to kill but to confuse." An officer interrupted to the surprise of everyone: "Sir, how do we get them close enough"? "Good question. You didn't think I would leave that out, did you, Lieutenant?" "No, Sir. My error."

Captain Jefferies continued, "Here is how we are going to do this. At 0400 hours tomorrow, we will detonate 10 of those mines a few seconds apart. At the same time, I want 200 of our fast men and women to start yelling and screaming while running towards the commies. They should be awake and ready to fight by the time our troops get into the enemy's visual range. As soon as our troops could see the enemy, they should stop, one or two of them should shout, "We are outnumbered," and turn and run back to this position. The commies are so desperate for a victory that they will chase a smaller force. The two hundred will lead them right into our killing zone. I will be in a tree and will pop a flare to signal the attack once I can see the enemy is fully in the trap. When the flare pops, I want 50 bombs thrown out. Some of the commies may even kill each other in the confusion.

When the bombs stop exploding, the rest of us will attack with knives and bayonets and fight until the enemy is eliminated. It will be a low-light level mission, so know your targets. I want to use the darkness to conceal troops waiting in ambush. If I time this just right, the sun will start illuminating the battlefield just as the heavy fighting starts. The Claymores will be deployed at the ends of our formation to kill any commies who try to run away. I want duty rosters for the six OPs and a roving patrol that started at dark. Grab your gear, get your people ready, and let's do this."

Everyone dispersed back to their position in the line except Sgt. Lowe. He was called for a private meeting with Captain Jefferies. He was told that he and his squad were to lead the two hundred troops and that he should take his dog with him just in case there were more mines. Sgt. Lowe asked the captain how he came up with such a plan, and he said he was inspired by some battles in the Bible and his son's Nerf football. Both chuckled a little, and Sgt. Lowe went back to his squad and told them the plan. They seemed optimistic but disappointed that the fight would not be today. It meant another night of tenseness and uncertainty as they lay, unable to sleep on the grass all night. Troops worked on the bombs and took up their positions for the first couple of hours, and then everyone just lay in place. There was no food and very little water, but there were also no complaints either.

Sgt. Lowe accepted the fact that the dog was his and he toiled with names for several hours. In the end, he decided to just call her a girl because she was clearly a well-trained bomb dog. That meant she belonged to somebody, and that there was a chip in her with her name and who she belonged to. She was an American dog, he figured, because she responded to English commands. He rubbed her head as he promised her that he would find out who she was and get her home.

Most of the soldiers were able to get a little sleep despite the gravity of what was about to happen. At 03:45 hours, though, everyone was awake and alert. The two hundred runners were ready, and only time was stopping the detonation of the ten mines that would start the mission. A heavy stick or rock was dangling over the prongs of each mine with a string. Soldiers only needed to pull the string at the other end a safe distance away to set off the mines. The tension rose with every second that passed on the clock. Then, at exactly 0400 Hours, the first mine detonated, followed closely by the other nine. The two hundred yelled and screamed as if death were chasing them. They moved at a quick pace towards the enemy, but not too quickly, as they needed to conserve energy for the run back. As they approached the enemy, they could hear them forming up for combat. Several commands were being shouted in German and French. The two hundred slowed up as they came within the view of the enemy. Sgt. Lowe could only guess, but based on his experience with bayonet battles, there were about five thousand of them. They did not have bayonets because Russia had long ago stopped manufacturing rifles with bayonet studs.

Most military leaders were convinced that the days of bayonet charges were forever gone. It was common to see an enemy soldier with a knife taped or tied to the end of his rifle. Some became quite effective with it.

Finally, Sgt. Lowe noticed an enemy soldier pointing at them. There was a moment of silence in the shouting from the enemy camp. Sgt. Lowe waited for the noise to die down a little, then he shouted as loud as he could, "We're outnumbered, retreat." Other soldiers repeated the retreat order, and soon jubilation could be heard coming from the other side. Sgt. Lowe knew Captain Jefferies was right; they were going to chase. Sgt. Lowe ordered his team to start the long run back, and they moved out at a much faster pace than they used on the way to the enemy.

The communist troops were following, shouting as they pursued. It was then that Sgt. Lowe found the flaw with this plan. The enemy was critically low on ammunition, but they were not out. The sound of AK-47s being fired filled the air. It was only for a few seconds, but the bullets found some of the Americans. Sgt. Lowe heard Pvt. Johnson scream out in agony as a bullet pierced his back and shattered a part of his spine. Sgt Lowe looked back at Pvt. Johnson and could see the pain in his now flushed face. He rushed to him and threw him across his shoulders, and carried him past the friendly line. As he ran, he could hear others crying out for help. Sgt. Lowe laid Johnson on the ground and ran back towards the wounded, encouraging others to follow him. The gunfire had subsided now, but the enemy was within a few steps of slaughtering the wounded with rifle butts and knives.

Sgt Lowe ran to the wounded soldier who was further away because he was closest to the enemy. He reached down for the fallen soldier when he caught the glimpse of a rifle butt just in time to move out of the way. Sgt Lowe fell in his effort to dodge the strike and the enemy soldier recoiled quickly and was now bringing the rifle butt down on Sgt. Lowe’s face. Sgt. Lowe could see the anger and determination in his adversary’s eyes as he waited for the impact. But the eyes turned large and showed fear in a split second. Blood fell on Sgt. Lowe's face in place of the butt of a rifle. Sgt. Lowe could hear growling as the enemy soldier began to scream out. The dog had bitten the soldier on the neck, tearing and ripping out a portion of his jugular vein. He bled to death in seconds.

Sgt. Lowe regained his composure by telling himself to get his head out of his fourth point of contact, picked up the wounded soldier, and carried him to safety. There were now medics at the location, giving aid to the wounded.

Sgt. Lowe laid the soldier down, patted the dog on the head, and returned to get another soldier. He carried out one more without incident and returned for another. He saw a wounded soldier and ran towards him. He could hear the bombs exploding, and he smiled as he knew the plan was working. He reached the wounded man just a split second behind an enemy soldier who had a rifle with a knife taped to the barrel held above the man's chest, and was a millisecond from thrusting the knife forward. Sgt Lowe screamed "No!" and dived on the soldier just as the enemy brought the weapon downward. Sgt Lowe's body went stiff in surprise as he felt the knife go through his back. The warm sensation of large amounts of freshly spilled blood overtook him for just a second as the enemy soldier withdrew the knife from his back and held it up to thrust into the chest of the dog. The dog was leaping towards the enemy, but he managed to get the knife around just in time. The dog's growls turned into a short, high-pitched whine as the knife broke a rib and penetrated the heart. Sgt. Lowe's surprise instantly turned to anger as he leapt to his feet and cursed the enemy soldier. The soldier charged Sgt. Lowe, with the knife leading the way. Sgt. Lowe found that he could only notice a few of the events happening around him. He could hear the dog whimpering as it struggled to breathe, the wounded soldier yelling, "I'm coming, Sarge," as he struggled to get up and help, the blood now running down his leg, and the knife fast approaching to finish him off.

Sgt. Lowe reached out with his left hand and caught the barrel of the AK-47 while stepping slightly to the right. He pulled the knife past him and caused his opponent to become off balance as he was pulled in front of Sgt. Lowe, his right shoulder leading, and his back slightly exposed to Sgt. Lowe.

Sgt. Lowe wrapped his right arm around the enemy soldier's neck, released the rifle, put his left hand across the enemy's forehead, and fell backward to the ground, squeezing with all the strength he could summon from his weakening body. The enemy soldier's ability to breathe was stopped by Sgt. Lowe. He kicked and punched, but the vice would not open. Death came in a few minutes and, only then, was he released. Sgt. Lowe threw the wounded soldier over one shoulder and the dog over the other. He looked down at his rifle and felt sadness as, for the first time ever, he left it on the battlefield. He made his way to the makeshift aid station, stumbling as he walked the last few steps. He laid the wounded soldier down, walked a few steps away, and collapsed. He rolled over to his left side, pulled the dog tight against him, and patted it on the head. "Good girl," were his last words. The dog let out a final whimper as it licked Sgt. Lowe's face. Sgt. Lowe's mind focused on the memory of a picture he once had of his wife and baby daughter. His wife was holding the baby overhead, and both were looking at the camera.

Sgt. Lowe could hear Captain Jefferies command his troops to move on to the enemy TOC and take it. He relished in the accomplishment as he and the dog exhaled simultaneously for the last time.

Chapter 8
An Unexpected Leader

There was a loud thud as Dr. Smitherson awakened while falling to the floor. The sound of items from his desk could be heard as well. He knocked the items off the desk when he tried to save himself from falling. Landing next to him was the service record of Sergeant Carr Lowe. Accounts of his last fight were recorded in the file. He was, by all accounts, a hero. Dr. Smitherson fell asleep just after reading about the actions of Sgt. Lowe and his dog were during the battle, and the power of dreams brought it to life for him.

Dr. Smitherson also received Sgt. Lowe's personal effects. The Army had no one to send them to because their entire family was killed in Seattle, Washington. His father, mother, brothers, and two sisters all died of radiation poisoning. His wife and young daughter were both killed in the initial blast of a nuclear weapon. In the box of personal effects was Sgt. Lowe's diary. In it, Dr. Smitherson learned that Sgt. Lowe was named Carr by his father because he was born in the car on the way to the hospital. His mother insisted on the second "R", so it did not seem so obvious. She was afraid that most people would ridicule her boy over the name, but it turned out that people liked it and the story behind the name.

Dr. Smitherson also read about how Sgt. Lowe would have the same nightmare several times a week. He was lying next to his wife in a park. They had gone there for a picnic, and they were lying on their backs in the grass. He was holding his baby daughter above him, and they were all smiling and laughing. Suddenly, his daughter turned to ashes and crashed down on his face and chest. He screamed as the last bit of ashes fell through his fingers. He turned to his wife only to find that she, too, was now ashes, lying on the bare ground, still in the shape of her body. All the grass and trees were gone, and he was the only living thing as far as his eyes could see. Sgt. Lowe wrote that he would awake in a cold sweat and with renewed anger that he was sent to New York instead of California, so he could kill those responsible.

Dr. Smitherson also found a Congressional Medal of Honor and a Silver Star in the box. The Medal of Honor was awarded to Sgt. Lowe, but there was no one to present it to. The Silver Star was awarded to the dog. The dog was indeed a bomb dog. Her name was Iron Side after the ship. Her former handler was killed in action, and it was assumed that Iron Side died as well.

At the request of Captain Jefferies, Sgt. Lowe was officially named as her handler, and that was how her medal ended up with Sgt. Lowe's personal effects.

Dr. Smitherson concluded that after the death of his family, Sgt. Lowe's only reason for living was to kill those responsible. He became more dedicated to the Army and soon was able to realize that winning the war was one way to get back at the Muslims. He always believed his day would come. Dr. Smitherson realized he had given Sgt. Lowe had a chance to get some revenge, and he wondered if Sgt. Lowe knew it. He did so well at the camp in California. He remembered things that no other RAS could recall, or at least did not act like they could recall. His ability to perform as a living, thinking soldier was most unexpected. Dr. Smitherson knew he had to find out what made him different. Why was so much of his memory saved in the brain stem when so little was saved in others? Dr. Smitherson pondered this question, struggling to find a researchable answer, while he got dressed for the weapons demonstration and later, his press conference. He placed a call to General Lancer and told him to contact Captain Jefferies. Dr. Smitherson wanted to question him about what happened to Sgt. Lowe's remains immediately after his death. He also wanted to find out what happened to the dog. He thought he might in some way be able to let Sgt. Lowe know. It would also be a good test for Sgt. Lowe's memory retention. He only knew the dog for a little over 24 hours. He had one other test in mind, but he would have to consult with Sam and Ben and get their approval before he could make it happen. He was also processing the next combat missions. It was becoming increasingly clear that two missions were going to be carried out on different sides of the country, and at the same time. Dr. Smitherson was starting to feel a little overwhelmed that he could not make it all happen. He needed some help, and he wished he could get advice from his brother.

Dr. Smitherson called the Manhart brothers and told them to meet in his office to discuss and experiment with them. After some thought, he decided his entire team should be involved, so he called the medical team as well. Soon, everyone was gathered in Dr. Smitherson's office. He knew there was not a great deal of time, so he did not beat around the bush. He asked Dr. Colman if number Thirty-one had been fitted with the safety device. She told him no, but it could be done in just a few minutes. There was no prep or recovery time. Dr. Smitherson then explained that he would like to use number Thirty-one in the weapons experiment. He felt that it was important to prove that it would work on an RAS trooper and not just a cadaver arm. Sam seemed very excited about the idea, but admitted he did not expect this, and he was not sure of the outcome. Thirty-one could turn and point the weapon at the observers, he explained. Dr. Smitherson told the team that the safety device was for. He went on to explain that he did not believe it would be an issue. When Dr. Chi asked him why, Dr. Smitherson told him that he, Ben, and Sam were about to go talk to him. He instructed the medical team to get ready to install the safety device and repair any injuries on Thirty-one, and he led Ben and Sam to the cooler where Thirty-one was being stored.

Number Thirty-one was strapped to one of the chairs in the long rows and started moving as he warmed up. His face slowly distorted to an evil look as more and more he realized he could move. Soon, it became clear that he was not as in control as Dr. Smitherson thought. Number Thirty-one wanted to kill the team and probably wanted to eat them. Dr. Smitherson began mumbling something about putting him back in the freezer and going with the cadaver arm when Dr. Colman ran over to them from the lab.

"I know what is wrong and I have prepared for it," she said as she got close to the group.

"What?" asks Dr. Smitherson.

"He is hungry. Not like when you or I get hungry. He needs energy that can only be found in human flesh." Dr. Smitherson was disgusted, but he knew she was right.

"What is your plan?" he asked.

"I have thawed one of the enemy soldiers killed in California, and he still has low levels of electricity that were trapped when he was frozen. It won't last long, so if we are going to do this, we need to do it now," said Dr. Colman.

Dr. Smitherson called for a moving team, and number Thirty-one was moved to a room pointed out by Dr. Colman. Number Thirty-one was pushed through the doorway with the bite pole, and a soldier carefully removed the leather strap. The strap had been in place so long that, at first, Thirty-one's cheeks seemed disfigured as they retained the shape created by the strap. The binders on his hands were cut, and the soldier ran as he was nearly grabbed by Thirty-one. A button on the bite pole was depressed and Thirty-one was released into the room, and the door was slammed shut. The team watched on monitors as Thirty-one circled the table with the corpse on it. He suddenly stopped and began sniffing the air. He followed a scent to the head of the corpse and immediately began tearing the head apart, devouring the brain as quickly as possible, being careful to consume every morsel. He then poured over the remains with his nose, searching for more usable flesh. When he could not find any, he walked to a corner and sat down. At first, he looked satisfied and content, but after a couple of minutes, he appeared to go into a kind of rest mode. He sat in the corner with his head resting where the two walls meet. His feet were drawn in close, his knees were raised, and his arms rested on his knees. His hands extended past his knees, and his fingers hung down. He shifted his eyes from side to side and up and down as if he were searching for something. And soon closed them.

A team of soldiers equipped with thick leather over garments and a bite pole slowly open the door and enter the room. Number Thirty-one's eyes opened almost immediately, and he rose to his feet. The soldiers were surprised by the distorted, evil-looking face of Thirty-one. It was as if he could see through them and could only focus on the food in front of him. His head was slightly tilted, and his eyes were narrow but clearly focused on the group of soldiers. His nostrils flared rapidly as the smell of the electricity inside the skulls was detected. One of the soldiers snickered as he remembered an old cartoon. A small chicken was walking past a hungry cat, and the cat saw the chicken as already plucked and cooked, with steam rising from where the head should be. The soldier wondered if the zombie only saw him as a giant brain or slab of meat.

Number Thirty-one now appeared to be looking around the room for weapons and planning an attack. The few charged neurons in the dead soldier would not be enough; he was using his reserve energy now, and he must have more. When the reserve ran out, he ceased to function forever. The soldier could tell that Number Thirty-one was getting ready to attack. Those with bite poles got a better grip on the handle, and all of them checked and rechecked the fasteners on their leather armor.

Doctor Smitherson broke through the group of soldiers and shouted, "Sergeant Lowe. Stand down." Number Thirty-one's facial expression changed to obvious fear for just a second, then back to evil as he wondered what had just happened. "I said stand down, Sergeant," he shouted again. Number Thirty-one froze and, after a few seconds of trying to determine what was happening, he said in a rough voice, "Yes, Sir". Doctor Smitherson walked right up to Thirty-one with no armor or defensive weapon and began talking to him.

"You are Sgt. Lowe?"

"I think so."

"Are you in control of your actions?"

"I am trying, Sir, but I am so hungry. What has happened to me?"

"I do not beat around the bush, Sergeant, so I will just lay it all out there. You were killed in action at the Battle of Lockport. Do you remember that battle?"

"I do remember the battle, but I don't remember dying."

His voice was getting weak and raspy, and his face showed unmistakable expressions of grief and a lack of understanding. He asked for a drink so that he could continue to speak, adding that it was difficult to force air over his vocal cords. "It is hard to breathe," he added, "when you don't have to." Dr. Smitherson knew that what was really happening was that the ability of his lungs to expand and deflate was degrading as his lungs dried out. He made a mental note to find a way to keep this from happening.

Dr. Smitherson asked for a glass of water and took it to Thirty-one. He was well within Thirty-one's killing radius before he realized he was so close. He extended his arm with the water and said, "Here, Sergeant." Thirty-one, who was about ten inches taller than the doctor, looked down at the crown of Dr. Smitherson's head. His nostrils flared as he forced air past them. His face reverted to the horrifying look seen before, just before a kill. His mouth opened, and he leaned slightly forward, his teeth pushed as far forward as possible, and ready to tear flesh and break bone. Panic began to grow in those watching the events unfold. A soldier pulled out the device that, if activated, would destroy Thirty-one's brain stem. He was about to depress the button when Thirty-one began shaking his head from side to side while moaning in disgust. He stopped and took the glass of water, saying, "Thank you, Sir," and he drank it down. "You're welcome," said Dr. Smitherson, who never realized he was in any danger.

"How is Pvt. Johnson?" Thirty-one asked.

"He recovered well, Sergeant. You saved his life along with two other soldiers."

"What happened to me? I can't recall dying. The last thing I remember was being very cold."

Dr. Smitherson remembered reading that Sgt. Lowe and the dog were placed in the enemy's freezer found at the TOC. There were a multitude of American injuries, but Sgt. Lowe and the dog were the only fatalities. Sgt. Lowe's remains were put on the first plane out and taken straight to Dr. Smitherson. The dog was preserved because Captain Jefferies thought that it should be mounted and displayed in the Army Museum for its heroics in battle. It was just a dog to him, but a dog that saved the lives of many of his men and women.

"What is the first thing you remember after the battle?" asked Dr. Smitherson.

"Waking up on a plane with an I.V. and a bunch of wounded soldiers. We were air-dropped over an enemy compound with no weapons and bad rigging. Most of my straps were missing. I think we won, though, and I ate." He paused for a moment, and he looked perplexed. "People," he added very slowly as if he was convincing himself at the same time he said it.

Dr. Smitherson started to explain, but Thirty-one quickly interrupted him. He explained that, for now, he didn't care what happened. He was hungry, and he had to eat. The weak power from the corpse on the table was not enough, and the fear of running out of reserve was taking control. To everyone's amazement, Thirty-one recommended he be gagged and his hands bound while someone figured out how to get him a live human. Dr. Smitherson had Thirty-one bound and shackled to a chair while he called General Lancer. Dr. Smitherson was looking for a close enemy target. He wanted something small yet important that Thirty-one could accomplish by himself, and it could be done in just a couple of hours.

General Lancer had the perfect mission. He was planning a Delta Force raid on a farmhouse believed to have a virtual treasure trove of intelligence documents. The house was only being guarded and maintained by five people at a time to help avoid the interest of the U.S. Military. There were also no civilians around for more than a mile. It was believed that the Muslim answer to RAS troops was currently in the house, making it an even more valuable target. General Lancer's new plan was to put Thirty-one in the home to kill the guards and then have Delta Force mop up and capture the intelligence documents. The plan called for an unarmed Black Hawk to drop Thirty-one over the farm from a high altitude because utility helicopters were a common sight in the area.

Everything was thrown together quickly because Dr. Smitherson wanted Thirty-one back in time for the weapons demonstration. He told Sam to prepare a half-hour speech to buy time and have it ready in case the mission ran late. Dr. Smitherson postponed his press conference indefinitely, using the need for his presence at some upcoming experiments as his reason.

Two Blackhawks arrived at the air base in a matter of minutes. One for Thirty-one and the other for the special ops unit. There was no time to plan for cameras to record the attack, but Dr. Smitherson believed Thirty-one would be able to tell him about it when he returned. The helicopters left at 0915 hours, and the weapons demonstration was scheduled for

1300. Dr. Smitherson was a little worried about the time and how Thirty-one would do on the mission. For a fleeting second, he thought that maybe he had put too much confidence in a corpse.

The Blackhawks were approaching the farm, and the pilot ordered the crew chief to take the leather gag and bindings off Thirty-one so he could prepare for the jump. Thirty-one was still unaware of the safety device in his head. He adjusted the straps on his rigging but looked to the crew like he was fighting himself to do so. He was fighting himself. All he wanted to do was rip the crew's skulls open and scoop out the brains like ice cream from a box. He could feel the smell of their energy moving up his nostrils and to his brain stem. It was exhilarating as his energy reserves were released, so he had the strength to kill and feed. He squeezed his nostrils shut to try to dampen the smell, and one of the crew members laughed at him. "What's wrong, zombie man? My bird doesn't smell good enough for you, you stinking pus bag?"

Thirty-one looked at him through narrow eyes. Anger could be seen despite the decay; the corners of his mouth drew back, and his lips seemed to disappear. His discolored teeth were as prominent as a wolf's fangs when defending her young. He slowly began leaning towards the crewman and, in a matter of seconds, was up on his feet and grasping the crewman in his hands. The crewman was the soldier carrying the remote for Thirty-one's safety device, and he pulled it from around his neck. Thirty-one quickly knocked it from his hand, breaking the ball chain that held it around the crewman's neck.

It fell to the floor and slid to the back of the helicopter and was out of reach.

Thirty-one quickly wrestled the crewman to the ground as the other crew members drew their side arms and pointed them at Thirty-one. No one fired a shot; half for fear of shooting down the helicopter and half because they believed the crewman was getting what he deserved.

Thirty-one grabbed the crewman's chin in one hand and forehead in the other. The crewman, by now, had given up and was trembling and pleading for his life. Thirty-one opened his mouth and lowered his head towards the head of the crewman very slowly and with a deep, ominous but quiet laugh. The crewman recognized the smell as the same smell in the air when he had to spend two days in a field of decomposing bodies. It was a stench of death that caused his stomach to contract and force the food stored in it up and out of his mouth. Thirty-one ripped off the crewman's communication helmet and stopped moving with his mouth a millimeter from the crewman's right ear.

"Mind your tongue, boy, or I will rip it out and eat it. Are we clear?"

"Yeeee, yes, sir," came the quivering reply.

Just then an alarm sounded and a yellow light in the back came on. Thirty-one crawled to the back and picked up the safety device. He stared at it for a few minutes, moving it around so he could examine it all.

"Two minutes," came over the intercom. Thirty-one put the strange device in his pocket and stood up. He adjusted the straps on his rigging and was thankful they were all there. A green light came on, and a crewman opened the door. Cold air rushed in as Thirty-one ran to the opening and jumped out. He knew he had an altimeter to open his chute, so he just thought of the impending feast. He was tired and low on energy. He fell for about 45 seconds, watching with amazement as the ground rushed towards him.

He then heard the fasteners release and the parachute being ripped from the pack on his back. He looked up as the canopy formed above his head and his descent slowed. For just a moment, he remembered Ft. Benning and his first jump at Airborne School. How free he felt, and how he knew immediately that he had made the right decision to join the Rangers.

The memory quickly faded as gunfire broke the silence. He could feel the impact as bullets hit him twice on his legs, but it did not hurt. He looked up and saw holes torn in his canopy, and his rate of descent increased quickly. This made him happy because he was getting close enough to the ground now that he could smell the enemy. He looked down past his feet and could see that one of the enemy soldiers had taken up a standing position directly below him and was shooting up at him. He could hear bullets buzz past him and felt a couple of impacts on his lower body. The soldier firing realized the stories about undead soldiers were true, and his heart rate increased as he realized he was now foolishly standing beneath one. He began to shout in Arabic as he held the trigger down on his AK-47 until the magazine was empty. He threw the rifle down and decided to flee when a boot crashed into his head, and the impact broke his neck. He fell to the grass, conscious and alert, but unable to move.

His eyes shifted from a close-up view of blades of green grass to the torn and tattered face of 31. Back and forth, his eyes shift, and the face gets a little closer each time. He wanted to scream, but only the sound of liquid being moved by forced air came from his mouth. In just a very few seconds, he can see the chin and feel teeth sink into his forehead. He could feel the agony as his skin stretched until it tore, as the chin began to move away. He saw the face come into view again just as the piece of flesh disappeared into the mouth, and then the neck expanded and contracted as his flesh was swallowed.

Then the process repeated, only this time it was fast, and he could feel the teeth slam into the exposed portion of his skull. There was a quick, hard bite followed by a loud crack, and the skull broke. He hoped for death now. The pain was unbearable. He could feel life fleeing him as his blood poured out on the ground. Thirty-one used his fingers to begin ripping the skull apart. The soldier couldn't believe he was still alive as fragments of bone broke away and more flesh was ripped away. Finally, Thirty-one drove his index and middle finger deep into the soldier's brain and scooped a large section and ate it. The soldier's vision faded, the pain subsided, and death finally came for him. He smiled as he felt the pain leave him.

Thirty-one grew stronger and more alert as he consumed the entire brain. He became aware of the bullets impacting in and around him. He could feel his head being thrust forward as small pieces of lead encased in copper slammed into the back of his helmet, some penetrating and getting trapped in the layers of Kevlar, skin, flesh, and titanium. He looked back at the house over his right shoulder, where the gunfire was coming from. From inside the house, three men could see Thirty-one still kneeling by their friend, who lay dead on the ground. The top of his head was nearly gone, and an empty cavity was in partial view. Thirty-one's eyes were narrow, dark, and piercing. Only his upper lip could be seen from his mouth over his shoulder, and yet the appearance caused such fear to come upon others that one of the men threw down his rifle and ran outside screaming in Arabic.

Thirty-one saw the man exit the house, and his nostrils flared as the scent of more food reached him. He jumped to his feet as if he was going to chase him, but suddenly stopped. He shook his head for a second, like when a person tries to shake out the lingering images of a bad dream, then picked up the dead soldier's rifle. He grabbed a magazine from the soldier, locked and loaded, then shot the soldier running away.

The bullet pierced the soldier's heart, killing him instantly. Thirty-one could now hear rotary wings beating the wind, and he knew he had little time to complete his mission. He ran into the door of the home, causing its hinges to dislodge from the wall. By now, enemy fire slowed in its intensity as ammo became low.

The two men left in the home were hiding behind an upended table, and each had only about 15 rounds of ammunition left. They were firing a shot or two at any sound they heard. Thirty-one picked up their scent as soon as he entered the house. He planned a hasty but careful assault, smiled, and charged the enemy position. His bullets ripped through the table and entered the body of one of the men. He screamed as blood exited him through a hole in his chest. The other man stood and begged for mercy as his rifle was dropped to the ground. Thirty-one approached the man quickly and, in an eerie but firm voice, told the man he would live and not be eaten if he turned over the intelligence documents. The man shouted no in English and mumbled some Arabic words. Thirty-one grabbed the man by his hair and tilted his head to one side. He opened his mouth and pulled his lips away from his teeth. The soldier blinked, and a picture formed in his head. He could see that Thirty-one was something other than human. His friend's hair and pieces of brain clung to Thirty-one's teeth.

The man shouts, "O Kay O Kay, I'll get them," he said with no accent at all. He led Thirty-one to a room with stacks of papers spread over three desks. There were maps and satellite images on the walls and mixed in amongst the documents. The man pleaded with Thirty-one to let him go, like he promised. He was crying and begging as Thirty-one grabbed hold of the man's throat and ripped a fistful of flesh out. The man put his hands around his neck in a vain effort to stop the bleeding. He gasped for breath as he stared at Thirty-one in total bewilderment. "I never promised, and I hate crybabies," Thirty-one coldly stated.

Thirty-one found several bags and put all the documents, photographs, and maps into them. The helicopters landed in the yard, and he could hear the special ops team approaching. He walked towards them, handing them four bags full of information, and said this was it as he kept walking past with an empty bag in his hand. As the team searched for more documents, Thirty-one went to each of the three dead soldiers, broke their skull open with the butt of his rifle and removed the brain. He placed the brains in the empty bag and walked towards the row of helicopters. The same crew that brought him here was taking him back. He handed the bag to the crewman, who mocked him on the way in and told him to put the bag somewhere cool, preferably ice. The crewman placed the bag under a seat, opened the first aid kit, and removed the three ice packs. He broke the vial inside each one and laid them around the bag. Thirty-one told him to put them inside the bag, and the crewman followed the instructions without comment. Smoke began to rise as the special ops team completed their sweep of the home. Their orders included putting the remains of any enemy soldier in the home and burning it to the ground. The special ops team members climbed into their assigned birds, and the flight took off towards the Air Base.

Back at his lab, Dr. Smitherson received a large package. Markings on the package read, "Danger dry ice." Dr. Smitherson had the package placed on an open autopsy table and had the security forces break it open. He watched closely, pointing out several hard plastic bladders marked liquid nitrogen, and told them to be very careful. He explained that marking the outside as dry ice made shipping the package less red tape burdened than admitting the use of nitrogen, but would still cause people to handle the package carefully. A large dog was left lying on the table as the wood and bladders were removed.

The animal was frozen, and Dr. Smitherson asked one of the guards to watch the dog and alert him personally as soon as it thawed. He added that the dog was not to be moved by anyone for any reason. It took several hours for the dog to thaw, and when it did, the guard on duty reported it to Dr. Smitherson. Dr. Smitherson called his team and had them meet him in the lab around the dog. Once the team was assembled, Dr. Smitherson shared his plan to reanimate the dog. He muzzled the animal and asked if there were any questions before he injected the Ichor. Ben asked why he was trying to reanimate an animal, and Dr. Smitherson explained that it was the dog that number Thirty-one died with. He wanted to see if the two would respond to each other and if animals responded better to reanimation than humans. A very brief discussion ensued, and it was determined that it would be in the best interest of the program to continue the experiment. The dog was strapped to the autopsy table, and an IV was started. Dr. Smitherson tried to inject the Ichor through the IV, but it was impossible because nothing was moving through the veins. He wondered for a moment how forceful the pump on the plane had to be to push Ichor through the veins of a hundred dead humans when he could not move it through one dog. He put a needle on the syringe and injected the Ichor into the dog's neck, near the base of the skull. There was no near-instant movement as with humans, but after 126 long seconds, the dog's limbs began to move. Soon, it was growling and its eyes opened. Dr. Smitherson had a large cage prepared for the dog, hoping it could be reanimated. The cage consisted of three walls made from concrete blocks, and the front was made from a heavy-gauge fence with a gate for a door. The cage was about 12 feet by 12 feet, and the walls and the fence went from floor to ceiling. He thought it would be better to move the dog as soon as possible before it became fully aware of its surroundings. The dog was placed in the cage, and the backstrap of the muzzle was unbuckled so the muzzle would fall off when the dog was released.

Dr. Smitherson bought some dog carcasses from the animal shelter and had pieces of flesh and brains in a bowl inside the cage. The dog began to kick its legs rapidly as it tried to get to its feet as soon as the door to the cage was closed. It found its footing and started running from wall to wall, crashing into the wall each time. The sixth crash slowed the dog down, and it paced for a few seconds, constantly growling as it walked. Dr. Smitherson caught the dog's eye for a second, and he was reminded of a scene from a werewolf movie. Anger and hatred were projected from the dog's face. For the first time since he started this project, he worried about the safety of his team.

The dog suddenly quit pacing and started sniffing in the air. His attention was directed to the bowl of dog remains, and he ran towards it, sliding to a stop with his head over the bowl. He took in one more sniff and started to slowly eat. His pace quickened with each bite. As the team watched, Dr. Smitherson pulled a pen and notebook from his shirt pocket and wrote himself a note to find out how reanimated beings can sniff when their lungs are not functioning.

The dog finished eating and walked to the front of the cage. She peered through the fence as if she were staring down at each person in front of the cage, growling as she moved her head from side to side. Her growls, though, turned to unnatural barks when she seemed to recognize a military uniform being worn by one of the soldiers in the back of the group. Dr. Smitherson motioned for people to move back and for the soldier to move forward. The soldier reluctantly stepped forward, his right hand firmly grasping the grips on his sidearm. "Go, go, good dog," he said, fear causing him to stutter as he spoke.

The dog whimpered and sat down on her hind legs, tail wagging. The dog sniffed the air around the soldier and, as if she did not find the smell she wanted, walked to the back right corner of the cage and lay down. Her head was perched on her front paws. She closed her eyes and did not move anymore. Dr. Smitherson was so intrigued with what just happened that he wrote it all down despite knowing the whole thing was captured by both audio and video equipment.

Dr. Smitherson was told he had an important phone call by a security force, and he quickly left the area to answer it. On the phone was General Lancer reporting that Thirty-one was on his way back, that the mission was an unbelievable success, and that Thirty-one had the triggering device in his possession. Dr. Smitherson was worried, not because Thirty-one had the device, but more about how he was going to explain to Thirty-one what it was. His thoughts were interrupted by the sound of helicopters landing outside. He cleared his throat as he tried to pull together in his mind just what he was going to say when Thirty-one asked him about the safety device. He was stunned to see Thirty-one walking down the corridor with his hands loose and no gag in his mouth. He was walking with the crew from the helicopter, and they seemed to be talking about the mission. Thirty-one was carrying a bag in his left hand but using his right hand to demonstrate what he accomplished during the mission. It sounded like there were even some jokes and chuckles of laughter. As the group approached Dr. Smitherson, Thirty-one's facial expressions became serious, almost evil. He stopped walking and reached into his pocket to retrieve the safety device. He pulled it out and threw it towards Dr. Smitherson, saying, "Hey! You might need this." "I get really hungry sometimes." Dr. Smitherson felt his face going flush as he reached out to catch the device. He was even more embarrassed when he realized Thirty-one and the crews were all in deep, robust laughter.

Dr. Smitherson joined in the laughter as soon as he realized Thirty-one had just made a joke. He wondered as he laughed how this could have happened. Laughter meant happiness, and happiness was an emotion. Thirty-one was just supposed to kill the enemy until he could no longer function. And that was all, just emotionless killing and nothing more. The other 98 left did that perfectly. So, what made Thirty-one different? What made him so special? Dr. Smitherson pondered these questions as he laughed with the group.

Dr. Smitherson told the crewmen to go to the debriefing and asked Thirty-one if he would talk to him alone. As the crew started to walk away, the man who ridiculed Thirty-one said, "I'm sorry for what happened to the bird, but you do stink." Both smiled as the crew walked away. When the crew was out of sight, Thirty-one told Dr. Smitherson that it was a half-hearted joke about the safety device. He admitted that he was hardly in control when he felt his energy level depleting. He described it as a survival mechanism that, at its peak, cannot be overridden. Dr. Smitherson said he would always remember that and come up with a plan to keep Thirty-one from reaching that point in the future.

Dr. Smitherson then asked Thirty-one if he would be willing to participate in the weapons demonstration. He explained how the system works and that all he would have to do is stand in place and let the weapon do the work. "Yes," he said. Dr. Smitherson then asked Thirty-one what he would like to be called when he was addressed by others. Thirty-one appeared to be deep in thought for a few seconds, then he replied, "Sgt. Lowe will work. I know what I am. I saw enough movies to know that I am a Hollywood zombie. I have no soul. I believe it went to heaven to be with the wife and kid, but I really don't care. I know right from wrong, but I could kill you now just as easily as I killed those men today and think no more or less about it. Still, I am Sgt. Lowe. I have a sense of duty, but it can quickly become second to survival. I will do whatever I am sent to do if there is food in it for me."

"Understood," said Dr. Smitherson. "Is that what I think it is in that bag?" asked Dr. Smitherson.

"Dinner," Sgt. Lowe replied.

Dr. Smitherson called a guard to take the bag to the freezer. When the time came for the weapons demonstration, Sgt. Lowe was chained in place by his feet. He was gagged with a thick leather strap, his left arm was strapped to his waist, and the weapons system was on his right arm. His right wrist was attached to a chain that was attached to a metal pole on either side. This was a safety precaution so the weapon system would not lock on to the body heat of spectators behind him. "This is one way to stake out a sector of fire," Sgt Lowe thought to himself. Dr. Smitherson was standing behind Sgt. Lowe, Ben, and Sam were on either side of him. The crowd was especially excited because they were not told that a real reanimated soldier would be demonstrating the weapon.

Dr. Smitherson asked Sgt. Lowe, if he was ready, and, after he nodded his head, "Yes," told the witnesses to put on their hearing protection and watch the system do its work. A few seconds later, a green light appeared on the weapon system, and heated targets rose at varying distances down the firing lane. Sgt. Lowe struggled to hold back a reaction as his arm was moved around without his consent. The weapon system fired 30 rounds in less than 20 seconds and knocked down thirty targets. When the shooting was complete, Sgt. Lowe raised the system as far as the chains would allow, examining it in both amazement and approval. His lower lip was over his upper lip, and his eyes were as wide as Dr. Smitherson had ever seen them.

It was easy to tell Sgt. Lowe admired the system just by the way his eyes moved up and down it. After several seconds, his face turned serious as he nodded his head in satisfaction. He then snarled, exposing his teeth, and fought against the chains for the spectators like a good zombie. A bite pole was placed around his neck as he half-heartedly struggled against his captors and was led away.

Dr. Smitherson asked the Manhart brothers to meet later in the evening to discuss the weapons system. He told them to come with an open mind. Ben and Sam walked away in discussion about what 'open mind' meant and what Dr. Smitherson was up to now.

Chapter 9
A Zombie, A Dog, and An Old Friend

After the impressive weapons demonstration, Dr. Smitherson attended a briefing concerning the intelligence Sgt. Lowe captured. There was not only usable information on the plans Muslims had for the next few days, but there was also important information they gathered on the long-term plans of the Communists. The information on the Communist included important data about the day Sgt. Lowe was killed. Military Intelligence was desperate to know why there was such an expensive and well-maintained freezer on the battlefield when troops were hungry and out of ammunition. It turned out that a double agent between the Muslims and the Communists was more loyal to the Muslims. He provided evidence that showed the Communists were collecting specimens to create their own Reanimated Division. The soldiers at the battle were willing to fight to the death for a chance to be brought back as an invincible warrior, or at least that was what the propaganda told them they would be. Not a single enemy soldier made it to the freezer.

It was only used by Sgt. Lowe and his dog. The information also showed Kasmodod was desperate. He kept it hidden for a long time, but now American leaders knew his ability to wage war quickly faded. He was putting nearly all his resources into reanimation so he, too, could keep soldiers in the field longer.

There were also hundreds of spies and double agents sent out to get hold of Dr. Smitherson. The plan was to make him work for them to ensure the safety of his family. He took the danger seriously but still chuckled. When asked what was so funny, he pointed out that their military intelligence somehow missed that he does not have a family. Everyone laughed at that revelation.

The Information on the Muslims was a little more significant. Despite suffering daily casualties, their ranks were beginning to grow. He did not know it, but one of Kosmodad's problems was that his fighters were defecting in large numbers, and it was being reported as deaths. It would seem the human spirit needed a deity to put its faith in and give a hope of life after death.

The American religion would not work because there were so many of them, and the hatred for the American way of life was just too great. Nearly every maneuver Kasmodod made against the Muslims was successfully countered, and now the Americans knew why. He was being sold out by his people for the hope of an afterlife.

It was also learned that another nuclear weapon was nearly complete. It was being built in the subway system of New York City by Muslim sympathizers. The plan was very simple. As soon as it was completed, detonate it, sit back, and watch the devastation to both Americans and the Communist invaders. Allah would kill them by the hundreds of thousands, one document declared.

Perhaps the most important information of all was that all the Muslim leaders were going to meet in one place, and in just three days. According to documents, they believed the Americans would find out about the meeting and send in the Reanimated Division to kill them.

Their security plan was based entirely on the defense against the RAS. The plan was cruel by any other standards, but had long been used by Muslim nations. The plan was to keep all the military members and dignitaries safely inside the building and put women and children on the outside as human shields. If the RAS came, they would kill the women and children, then go into their rest mode. The soldiers would then come out and kill the RAS. There would also be an anti-American propaganda tape made because they would record the women and children being killed. General Lancer told his command team that he had an idea that might end the war, or at least America's involvement in it. He called them to his war room, and they began making plans.

Dr. Smitherson went back to Sgt Lowe. He was sitting on a bench with his feet shackled, but no other safety precautions were in place. He asked Sgt. Lowe if he was in control enough to accompany him to a surprise. Sgt. Lowe stated he was, and Dr. Smirtherson had his shackles removed, and they walked to a door that was new to Sgt. Lowe. He carefully examined it as he struggled with a deep urge to run. He was not afraid, but his survival instinct was starting to work on him.

His fears were soon replaced by elation as he passed through the door and turned a corner. There, in a cage, was an old familiar face. The first thing he could recognize since being reanimated. He looked at Dr. Smitherson and asked him if the dog was alive. Dr. Smitherson blushed as he replied, "In the same sense that you are."

Sgt. Lowe whistled the best he could, and the dog raised his head. He growled very briefly, then leaped to his feet with his tail wagging. Sgt Lowe opened the door and ran inside. The dog jumped into his arms, nearly knocking him down. The dog licked Sgt Lowe's face so emphatically that he had to say no for fear of his skin and flesh being pulled off. The reunion lasted over five minutes. During that time, everyone forgot they were watching a dangerous weapon on the monitor, and many of those watching were moved to tears. Dr. Smitherson allowed the dog to leave the cage. She believed that humans were not food interests to her, and she would only attack now if Sgt. Lowe told her to.

Dr. Smitherson told him that Sgt. Lowe wanted to talk to him "off the record," and that the only place in the building not monitored by audio and video recorders was his office. He realized the security and safety risk he had just taken after he said it, and he was not one to rethink a decision already made. The three made their way down a corridor and down onto a long tunnel. Sgt Lowe, who was keeping track of his location using a pace count, estimated the tunnel to be about 200 meters long. It was unpainted concrete and very uninviting. There were cameras in the tunnel, but Sgt. Lowe doubted there was enough light to make them effective. The light fixtures were spread so far apart that you were almost in darkness before you reached the light from the next fixture. The tunnel continued for what looked to be an indefinite distance, but the trio stopped at a door and passed through. It was like entering another world. The temperature was just right, and the walls were bare but neatly painted. Everything was neat and orderly. And most importantly, the smell of decaying flesh was gone.

Dr. Smitherson offered Sgt. Lowe a seat, but instead he and the dog walked over to the strange decoration in the corner. The dog barked and growled at it until Sgt. Lowe could convince her to stop. Sgt. Lowe stared at the brain stem, tilting his head from side to side as he got closer. He asked Dr. Smitherson how to open it and was told to just lift the glass case. Sgt. Lowe did so and then dropped the case almost as quickly as he lifted it. It was made from safety glass, so it did not shatter, but the noise startled everyone. Sgt. Lowe turned and looked at Dr. Smitherson with a seriousness on his face that Dr. Smitherson had never seen until now. "Your brother says hi," Sgt. Lowe said as if it was expected. He assumed it was the start of a daily conversation and did not know what to do as Dr. Smitherson's skin turned white and he stumbled around trying to maintain consciousness. Dr. Smitherson finally collapsed into his chair and took several slow, deep breaths.

"How did you know he is my brother?" he asked Sgt. Lowe

"He just told me," Sgt Lowe replied.

"How do you communicate with him?"

"Telepathically, I thought you knew that."

"I had no idea. I have so many questions. Will you ask him if he is up to a conversation, and will you translate for us?"

"He can understand you just fine, and you could understand him if you would try. I'll translate until you figure it out."

The conversation was odd by anyone's standards. Part thought only and part speech. Dr. Smitherson found that his brother was happy with what he did and even said he was sorry he tried to eat him. Dr. Smitherson also learned that one of the side effects of reanimation was telepathic communication.

He realized that the RAS troops were talking to each other using only their minds. That was why they would group. One or two would send a message that a new food source had been located. Others would then come to the food source until the signal went out that it was gone. They were bonding on a very primitive level. The conversation also revealed that Sgt. Lowe was much more capable than other RAS troops. He had control, to a degree, over his drive for food. He could talk and follow directions, too. The three weighed opinions and exchanged ideas until they all agreed on what was now the glaring reason. He was still alive when he was injected with the Ichor. He was clinically dead. No heart rate and no noticeable brain activity, but he was not completely gone. He was frozen quickly and, as the saying goes, "You're not frozen and dead until you're thawed and dead."

Dr. Smitherson gave Sgt. Lowe an apology for killing him, but Sgt. Lowe was quick to counter him by telling him he was dead, and, because of Dr. Smitherson, he could still fight. He still had a purpose. He made it clear that he was content with that and nothing more needed to be said.

Dr. Smitherson also learned Barny needed an energy source and soon. He told him that he was dying a second time. He needed so little energy to survive, but he did need it. Sgt. Lowe tried an experiment. He held his arm up to Dr. Smitherson's brother, exposing the holes that were made by the weapon system. He told him to try to guide his nerve endings into the holes to absorb some energy. It did not work. There were no muscles or tendons to guide the nerves. Moving them was impossible. They were too small for Sgt. Lowe to handle. Dr. Smitherson would have to inject a small amount of Sgt. Lowe's snack if his brother was going to survive. He would do so as soon as the conversation was over.

The three talked for over an hour. They all brought memories up from long ago. Sgt. Lowe talked about his time in the military and the various battles he fought. He shared his frustration that he never got to fight the Muslims until after he was dead. He talked about his wedding and the day his daughter was born. Dr. Smitherson was taken aback when he mentioned how much he missed them. He asked Sgt. Lowe if he still had feelings for them. Sgt. Lowe said his answer was complicated. He thought about them and would like to see them again, but he was not bothered by the fact that he never would. He was not sad, but the thoughts were still in his head. Dr. Smitherson admitted that it was complicated. He made up his mind not to study this phenomenon after the war.

Dr. Smitherson and his brother talked about their childhood and all the trouble they caused with their experiments. Both were very intelligent, and they had parents who supported them. They were always in first and second place at the science fair. They talked about how much they missed their parents. Dr. Smitherson was grieved by the thoughts, but Barny just missed them. There was no emotion, just the knowledge that someone was now gone. They also talked about how Dr. Smitherson was nearly eaten by Dr. Smitherson. The group all laughed as he described the look on his brother's face and the moment of panic he must have felt. Dr. Smitherson asked his brother if he was bored with what his existence was now. He said he was not. He would listen to conversations and laugh at people trying to guess what he was. When no one was around, he would conserve energy.

Sgt. Lowe demonstrated that telepathic communication worked with the dog as well. The dog would sit up, sit down, roll over, and shake hands without a single audible word being spoken. It was at this point that Dr. Smitherson realized no words had been spoken in quite some time.

He had been talking to his brother and Sgt. Lowe without talking. The other two laughed in their minds as Dr. Smitherson was filled with both pride and amazement. He wondered if he could communicate with all the RAS in the same way. New ideas of command and control, as well as healing traumatic brain injuries, flooded his mind. He waited for a response from the other two, but it never came. He then realized that it was not mind-reading but rather an actual form of communication. He had to project with his mind the words he wanted others to hear, just like he did with his vocal cords if he were speaking. Dr. Smitherson asked Sgt Lowe to go with him to get an injection ready for Barny. Sgt. Lowe followed, as did the dog. As they walked down the corridors, Dr. Smitherson asked Sgt. Lowe, "What is the dog's name?" Sgt. Lowe said he never got a chance to name it, and he really could not think of one. They both threw out names like Rover, Rex and Shep, but none of the standard dog names seemed to fit the dog. As they walked, they talked with both verbal and telepathic words, confusing those who were watching the security cameras. Sgt. Lowe sent the name Moose, and the dog began to bark. That was it then. The dog's new name was Moose.

The freezer door was opened, and Sgt. Lowe removed one of the brains from the bag. He held it out as Dr. Smitherson inserted the small needle of a syringe into the nearly frozen organ and pulled back the plunger handle. He withdrew a milliliter of fluid and small pieces of solid matter. He then withdrew the needle and recapped it. Sgt. Lowe then ate half of the brain and returned the other half to the bag. He waited for Dr. Smitherson to show a sign of disgust, but that never happened. Sgt Lowe wiped his lower face with his shirt sleeve, and the pair headed back to Dr. Smitherson's office.

Dr. Smitherson asked Barny if he was ready for the injection, and after a brief telepathic argument over the best injection site, he said he was ready. The glass case was opened, and Dr. Smitherson lifted the brain stem up gently in his left hand.

He then punctured the bottom with the needle and slowly depressed the plunger. He expected some of the fluid to leak as he withdrew the needle. Instead, the lining sealed itself as the needle was being pulled out.

Dr. Smitherson carefully placed the brain stem on the stand and was about to replace the glass case when he could hear in his mind, his brother screaming in agony, saying, 'It burns, it burns." Sgt Lowe rushed over to help, and both he and Dr. Smitherson were in an obvious state of panic. Then both picked up moans coming from Barny. It sounded like a bad actor in a "B" movie. Dr. Smitherson held his hands out in front of him and turned his head from side to side, trying to find something to do that might help. Suddenly, the telepathic calls stopped, and there was silence within the silence. Dr. Smitherson lowered his head and put his right hand on his forehead in disbelief. He just killed his brother was all he could think.

Just then, laughter filled the air inside the heads of Dr. Smitherson and Sgt. Lowe.

"I have been dying to get you back ever since you put that vampire mask on and jumped into my bed," Barny said.

"Give me a break. I was eleven years old," said Dr. Smitherson through loud and yet inaudible laughter.

"What did I do?" asked Sgt. Lowe

"Sorry. You are collateral damage," came the reply from Barny.

The three continued in conversation until Dr. Smitherson received a call from General Lancer. It was time to bring everyone to the table and develop a plan that would end the war. The President was expecting a plan in three hours.

As the team was assembling, assistants were busy preparing Ichor and dead soldiers for the next mission. The original 98 were in cold storage, and 6000 more volunteers lay in cooled-down warehouses throughout the base.

Chapter 10
Analyze. Plan. Prepare

General Lancer and his aides, intelligence officers, General Lincoln, who was now the commander of the First Reanimated Division, and all of Dr. Smitherson's team were in a large conference room with an eight-foot monitor hanging on every wall. There was a paperboard nameplate for each attendee as well as a microphone and keyboard. All the seats were occupied except for Dr. Smitherson's. People began to worry because he was never late. Voices rose to a low roar as speculation of where he was varied from person to person, and yet each person knew for sure that they had to be right. The roars quickly turned into stunned silence, except for two screams, when he entered the room with Sgt. Lowe, a zombie dog, and that quirky little model he kept in his office. General Lancer ordered security to restrain the RAS, but as they moved towards him, he told them to go get help. He then bared his teeth and roared at them. Their pace quickly slowed to a standstill as neither of the three wanted to be first to make contact. Sgt. Lowe and Dr. Smitherson started laughing, and that infuriated Gen. Lancer. He shouted over laughter and mumbling, demanding an explanation.

Dr. Smitherson pulled up a chair next to his for Sgt. Lowe then asked everyone to take their seats. He began an unrehearsed and unwritten speech.

"I know many of you have questions, and several are even frightened. Give me a few minutes and I will explain everything. This is Sgt Carr Lowe of the United States Army. He was killed at the Battle of Lockport a few days ago. With him is the dog that was killed as they saved hundreds of soldiers. I can tell by your facial expression alone that many of you are afraid. Your fear comes with good reason. You have seen or heard about the RAS video and how dangerous a RAS can be. Hopefully, what I tell you and demonstrate to you now will alleviate your fears as they pertain to Sgt. Lowe. Sgt. Lowe is a RAS, but he is much different than his counterparts. He is in control of his ability to kill and can rationally think and communicate his thoughts. The dog is also in control and follows the commands of Sgt. Lowe. We have developed a theory on how this happened, but we cannot share it until my medical team and I complete additional research. In the glass box on the table, there is no model. It is my brother, Dr. Barny Smitherson. He is still functioning, and we have found a way to communicate with him. His knowledge of my research and firsthand experience with Ichor will greatly assist in future research.

I know many of you must be thinking that I have lost my mind, so I have prepared a little demonstration that may not completely convince you, but it will certainly get your minds traveling in the right direction. The first step is to take all you know about human communication and set it aside for a few minutes." Dr. Smitherson then asked for a volunteer who did not know him personally but was willing to seriously help in the demonstration. He then instructed the volunteer, one of the soldiers providing security, to have the group come up with three sentences and write them down. One would be for Dr. Smitherson, one for Barny, and one for Sgt. Lowe. They would leave the room and go down the hall to ensure they did not hear the conversation. Once all three sentences were written down, the volunteers would call the three back into the room, and the volunteer would telepathically tell each one their sentence. Dr. Smitherson and Sgt. Lowe would then repeat the sentence. Barny would give his response to Sgt. Lowe, who would then tell the group.

The three left the room, with the dog in tow. The volunteer watched as the group disappeared around a corner at the end of the hall, then went back inside and told the group what they were to do. Five minutes passed just getting everyone to participate, and it took another fifteen minutes to get a consensus on three sentences. The three agreed on are:

My favorite pie is cherry.

Race cars go fast.

This whole thing is a shame.

The volunteer called the group back into the room and asked how he was to proceed. Dr. Smitherson told him to project Sgt. Lowe's question to him with his mind. "Think about speaking, but do not speak." The volunteer snickered a little but did as he was told. In less than a second, Sgt Lowe said out loud, "Race cars go fast." Dr. Smitherson then asked Sgt. Lowe to go into the hall. Once he was out of the room, the volunteer was told to project Barny's sentence in the same way.

There was no snickering this time. The volunteer knew he was not part of a hoax. He said he was finished, and Sgt. Lowe was called back into the room. Before Dr. Smitherson could speak Sgt. Lowe said, "His favorite pie was cherry back when he had a mouth." The words brought laughter and applause from some, but it was still clear that some still did not believe. Their facial expressions and headshakes portrayed doubt.

"I'm next," said Dr. Smitherson. The volunteer looked down at his notes to make sure he worded the sentence correctly. Dr. Smitherson quickly responded with "It is not a shame. This is not a show. We do not have time for a show." He then called General Lincoln forward and asked him directly if he believed what he had just seen. "No," he sharply replied. "I was told these things were weapons used like a rifle or grenade. "He started to say another sentence when Sgt. Lowe interrupted. "We are, Sir. We are indiscriminate in who we kill, and once released, we cannot be called back just like a bullet or a grenade. Dr. Smitherson is working on the best way to use the weapon. I do not understand your objection to that. His little demonstration was not to impress anyone or put on a show. It was to demonstrate a previously unknown feature of the weapon."

"But this is so difficult to understand or believe," came General Lincoln's response.

"I know," said Sgt. Lowe. Think of something from your life before the military. Something no one in this room could possibly know and tell me about without using your mouth or vocal cords."

Gen. Lincoln stood silent for a moment, then Sgt. Lowe said, "Being bit by a Northern Pike at eight years old would make me afraid of fish too." Gen. Lincoln did not say a word but the look of shock on his face told the audience that Sgt. Lowe was right.

The General's face turned pale, and he had to use the shoulders and chair backs of every person he passed to support himself as he walked back to his seat. He may not have seen a ghost, but he certainly just heard one.

Gen. Lancer then took over. "OK. Now that all of that is over, let's talk about what we need to do to win this war. We really need to take advantage of this data telling us that all the Muslim leaders are going to be in one building and kill them all. We know they are going to use women and children as human shields. I think that is to deter us from using the RAS. I think we should still use them. As cold as it sounds, the RAS does not care who they kill. What I think we need to do is drop them in two waves. The first will kill the women and children, and the second will kill the soldiers who come out to kill the first wave."

The sounds of disgusted people rose and interrupted Gen. Lancer. It seemed no one really liked his plan.

"I know you don't like it. I don't either. Someone tell me why it won't work or give a better solution, and, before God, I will change the plan." The murmurs turned into silence as the group realized there was no better solution to the problem. Some wept as they realized what their silence had just condoned. Gen. Lancer waited a full thirty seconds before Sgt. Lowe stood up to speak.

"Sir, I agree with your plan. Once I reach a point where I must have energy, I do not care who I kill. I may have a little more control and ability than my comrades, but make no mistake, I am an RAS trooper, and I will kill whoever I can to preserve my existence. My fellow RAS troopers will do the same. I am also sure that the enemy is not expecting a second wave. They will be killed, and we will win. I do think, however, there might be an equally effective way that will accomplish the objective and spare most, if not all, the women and children."

"I'm listening, troop," barked Gen. Lancer.

"Send in fake zombies," said Sgt. Lowe.

"What do you mean, sergeant?" asked the general.

"Get some volunteer Airborne and Rangers. Use make-up to make them look as much like me as possible. Put M-16s under their shirts and give them chloroform. They can use chloroform to put the civilians to sleep and pour some sort of fake blood and animal flesh on them. Once they are all unconscious, the soldiers can sit down and appear to be in resting mode. When the enemy comes out to kill them, our soldiers can kill them and then go kill everyone inside. The only risk to the women and children is stray rounds and a reaction to the chloroform. The soldiers would have to be aware of the camera locations and hide their actions while they "kill" the women and kids, but I think it can be pulled off. The element of surprise would be just as great as a second wave of RAS. Remember, they do not know about the weapon system designed for RAS troops, so guns will catch them completely off guard."

"We need to develop more details, but I like it. Take a break while I go call the president," said General Lancer. "That will free up the RAS for an attack against Lancegoroff's headquarters in New York. He clearly was the head of the Communist snake now, and if the head can be severed, the snake will die. "

General Lancer was gone for about ten minutes. Most of the group went after something to drink and to the restroom. Others came to talk to Sgt. Lowe. They wanted to conquer their fears and believed the room was a safe environment to do so. The General returned to the room and called for everyone's attention.

"The president has given his approval for Sgt. Lowe's attack plan. I have people gathering make-up artists and the 82nd has been recalled from the lines and is on their way here for briefings and training. Guns, ammo, and chloroform are also on the way.

The rangers are going to be used in coordinated attacks on Islamic bases in America and the Air Force and Navy are going to bomb targets in other countries. A SEAL team has been assigned the task of neutralizing the nuke and killing those responsible.

As for our Commie friends, they get the Marines, the Big Red 1, the 101st, and the RAS. All other military units will move forward wherever they are at the same time. Needless to say, nothing this big has ever been coordinated before. Timing is everything and every unit must be successful if it is going to work. We have been asked to find a way for RAS and the 101st to work together. If we cannot, there will be a coverage gap and Lancegoroff might find it and get through.

I understand Dr. Smitherson, and his team have some updates on the RAS. When they are finished, we will kick around some ideas about live soldiers working with reanimated soldiers and see what we can come up with. Dr. Smitherson." Dr. Smitherson and his team walked to the podium.

"First. I know you want to know how all of this is possible. I cannot share that with you because we are still researching to gather all the answers. What we do know is that Ichor only reanimates. It cannot keep a reanimated corpse alive. It seems that the only substance that can keep them alive is the flesh of the same species. In other words, only the flesh of a human can maintain a reanimated human. I realize how horrible that sounds, but remember, they are weapons. All weapons have horrible results on humans. Sgt. Lowe's dog has no interest in harming a human unless it is told to do so. If, however, you bring Fido to work, he could become a meal. I have asked Sgt. Lowe to elaborate more on this subject."

Dr. Smitherson took a step back and Sgt. Lowe stands at his seat. A look of uneasiness still showed on many of the audience as he started to speak.

"First, I want to tell you that you are all safe. I am not hungry. Secondly, I want you to know that I know what I am. I am a monster. That is what I would call myself if I were in your seat. I am also quite OK with that, so don't feel like you must act like you are my friend.

Now let's get into what drives an RAS. First, it is not the flesh they are after. It is the electricity produced by the human body. We must have that, or we will die. The human body uses electricity to make the heartbeat, tell muscles to move, and send messages throughout the body. We can get electricity from any part of the body if it is near the time a food source dies. The best place, though is from the brain.

It does not take much to satisfy a reanimated soldier but a natural, unbeatable drive to survive takes over when the power starts running low. We can no longer produce our own, so we must find it in someone else. Once the power levels are up to a point where there is no danger of running out, we naturally rest to conserve it. A RAS will not fight if there is no need for food. We are zombies and, though I have maintained my sense of loyalty, most do not care about a cause or what side of a war they are on.

We find food through our noses. We can see and hear but the sense of smell is the most effective tool. Our noses do not smell but rather detect usable electricity. Most cannot speak vocally but all are able to communicate telepathically. You all can do so as well but it takes practice. For a reanimated corpse, it is natural. I hope to be able to lead many of the RAS in battle using orders given telepathically, appealing to what may be left of the memory of military service.

Finally, we feel nothing. No pain whatsoever. I have had several rounds go right through me and thought nothing of it. We also cannot be destroyed except by destroying the brain stem. Severing the head of a RAS will only result in a living head still looking for food. Our existence centers on the brain stem. That is why Barny Smitherson is still "alive" in that glass box. He can communicate, has memory, and can still solve problems.

Lastly, we have no soul. We are just biological weapons. I can remember my family, and the thought of how they died is disturbing, but I have no feelings for them. They are dead and gone, and so am I. Think of it as buying a used computer. There may be information stored on the computer from the previous owner. It may even be family pictures or important financial documents. They are on the computer, but the computer does not care. It has a new life with its new owner.

I'll end by saying, 'Fear us.' We just want to eat. We are weapons. Use us like weapons and not like people."

Silence filled the room for a few seconds after Sgt. Lowe took his seat as the gravity of what was just said settled into the minds of the listeners. They all knew he was right, regardless of how "human" Sgt. Lowe made the RAS appear to be. Some even grew angry as they realized Dr. Smitherson had exposed them to such great danger. Dr. Smitherson then said his team was going to speak jointly on research and progress within the RAS program.

Sam and Ben went first. They described the weapon system as a wonderful success. Ben started trying to give a technical explanation of how it worked, but that fell flat. Sam simply said that when the system was armed, it controlled the RAS's arm and used electrical impulses to move the arm and align itself with a target. When the ammunition had all been expended, the weapon fell away and the RAS again had control of its arm. Most of the systems could be recovered and those that couldn't, wouldn't be remotely destroyed to prevent enemy reverse engineering.

Ben continued with the battle readiness of the First reanimated Division. It was at this point that Gen. Lincoln perked up and started taking notes.

We had 99 RAS who had been reanimated and were combat-ready. We had a total of 6081 volunteers in cold storage and enough Ichor to reanimate them when they were needed. Our contractor had 3000 weapons systems built, tested, and ready to go. They asked for questions but there were none.

The medical part of the team then stepped to the front. Dr. Coleman again spoke for the group. She started with some stomach-turning, but important facts. She explained that a reanimated corpse did not digest food. It used neurons to remove electricity from the flesh and tissue as the food was being chewed. The flesh and tissue then fell into the stomach or abdomen and rot. The decaying matter would need to be removed regularly if the corpse was going to be used multiple times.

The reanimated corpse was itself, decaying. We did not know how long a corpse would last but, from what we could measure, it appeared to be about three to five years without cold storage. The time frame was currently considered indefinite with cold storage.

As a survival mechanism, it produced one of the strongest insect repellents known to science. In tests, a fly would lay eggs on the side of a glass aquarium before laying them on a dead rat treated with repellent. This defense prevented insect larvae from consuming tissue. We were making good progress in research to make this available to live troops in the field.

She finished with a first-aid plan for RAS troops. Since they feel no pain and had no risk of infection, repairing them could be done with pieces of metal, super glue, and duct tape. When skin and flesh were torn, duct tape could be used until the tissue was repaired.

A broken bone could be exposed by pushing the flesh out of the way, applying a quick bonding glue to the ends, and setting them. Then in serious breaks, steel rods were taped to either side of the bone. The muscle tissue was then put back in place and the wound taped up. Better, longer-lasting repairs could be made in the lab after battles.

Dr. Jones spoke about tissue regeneration, "Progress was made that will allow a sample of a person's skin to be "grown" into a piece to cover serious wounds. This was not new technology but now it could be done in a matter of minutes as opposed to weeks or months. Very small amounts of Ichor are used in the process so research on living subjects has been limited. The process has worked very well on two RAS troopers. Large portions of skin were removed by flying debris, leaving gaping wounds that could not be glued or sewn back together. A sample of skin was taken from each and "patches" were grown in just a few minutes. The patches were laid over the open wounds. They bonded to the surrounding skin tissue and remained in place since. There has been one experiment on a living person. This soldier volunteered after her skin was peeled away from her legs in a motor vehicle accident. Her wounds were successfully covered, and the healing process went very well. She is still being monitored for side effects from the Ichor." Dr. Smitherson lowered his head in shame because he knew what Ichor would do to a living person. Kill them.

Dr. Chi then talked about the most exciting part of the report, at least to military commanders. "The medical team had already discovered that RAS troopers were attracted to the electric impulses generated inside a living person. As part of their goal to keep the RAS troops in the battle longer, they planned to train combat medics to make repairs to the RAS in the field. To solve the problem of the RAS attacking the medic, they designed a uniform that was laced with very thin copper wires, a small transmitter, and a battery pack. Very low-voltage electricity was sent through the system, disguising the wearer's electric output. A transmitter on the waste put this small current in the air, even further hiding the fact that the wearer was a living human and a food source from the RAS. The team had already tested the uniform by wearing one in a room with restrained RAS while they were in rest mode. They did not move. They stayed in rest mode, even when a wearer would pull up a chair and sit down next to an RAS. This could be the answer to how both living and reanimated soldiers could fight side by side."

General Lancer thanked everyone and dismissed the civilians. He now needed to weigh all his options and get a battle plan ready to present to the unit commanders when they arrived. He kept his aids with him, and they began debating on details like what unit would go where and who was going to go with the RAS. Time was so critical that it was decided to forgo additional testing on what was now being called the screening system.

Dr. Smitherson, Sgt Lowe, and Moose left together. Then Dr. Smitherson met Dr. Colman in the hallway and told her about how Sgt. Lowe most likely died. He was concerned for the woman with the new skin and asked to be taken to see her. She took them to her room where the doctors asked Sgt. Lowe to wait in the hall. They entered the room and found Specialist Joan Hunter sitting on her bed and clearly distressed.

"Everything alright, Joan?" asked Dr. Colman.

"No," she said. "I don't feel like I'm me anymore and I am so hungry that I think I might die if I don't get some food."

The two doctors looked at each other then slowly exited the room.

"Is she dead?" asked Dr. Colman.

"Yes. I think so but I don't think she knows it yet. She has already been reanimated. I don't think she even closed her eyes," said Dr. Smitherson, in more of an inquisitive tone than a compassionate tone.

"I know how she feels. I'll go talk to her, said Sgt. Lowe."

She looked up as he entered the room and jumped to her feet. She was not afraid but more concerned.

"Why are you in my room and who are you?"

"I am Sgt. Lowe of the First Reanimated Division. I don't beat around the bush, so I'll just put it out there. You died because of your injuries, and you have been reanimated."

"But I don't remember dying."

"Me either but we are both most certainly dead."

"What are you feeling right now?"

"I need to eat but I don't want food. I smelled something I wanted a few seconds ago."

"Stay here and I'll go get you something that will help."

Sgt. Lowe left the room and sent Dr. Smitherson to the cooler to get the bag he brought home from the farm. He put his brother, who he had been carrying the entire time, on the floor and headed off to the cooler. He opened the door and picked up the bag like anyone else would pick up a bag of fast food. He did not like the fact that his cargo was human brains; he was just indifferent to it. This sack of brains served no other purpose than to be food for two RAS troopers who needed them. He dialed seven digits on his cell phone as he was walking back and told a clerk to bring Spec. Hunter's military records to her room.

Dr. Smitherson arrived at the room to find Dr. Colman terrified and Sgt. Lowe was trying to restrain Spec Hunter against the wall. He quickly ran to the two RAS and held up the bag, so Hunter could get a scent of what was in it. She stopped struggling as he handed the bag to Sgt. Lowe and they both went back into the room. Inside the room, Hunter grabbed the brains and began eating like she had not eaten in days. As she ate, she calmed down, and her face transformed from looking like pure evil to looking almost like she did before she died. She quickly realized her new way of communication and Sgt. Lowe explained everything to her, including the fact that she was most likely killed by the treatment designed to save her legs. He went over the planned attack and the RAS's involvement in it.

As Sgt. Lowe was explaining the ways of zombie life to Hunter, and Dr. Smitherson and Dr. Coleman were reviewing her military records. She joined the Infantry as soon as that job was opened to women and fought in six battles. She had 17 confirmed kills and was at the air base for a stopover on her way to Drill Sergeant School. She was riding in a car hit by a dump truck, and the dashboard was crushed into her legs. Both vehicles caught fire, and it spread quickly. There was no time to wait for the rescue equipment. Bystanders pulled her from the car, leaving most of the skin on her legs as well as pieces of muscle tissue under the dash. It was noted in her file that she never screamed or became unconscious during the accident. There was one very serious problem. She was Catholic and not a volunteer for the RAS. She was told that traditional medicine would take weeks or months to heal her wounds, but the experimental method could take as little as a day or two and save Drill Sergeant School for her. She was also told that there were many possible side effects, including death, but she was not told that she might be reanimated. Both doctors decided that the only ethical thing to do was to terminate her immediately.

They waited for Sgt. Lowe to tell them it was safe to go into the room. Once inside, they began to explain what had happened and what their plan was. Hunter told them that she wanted to join the RAS now and that there was no need to terminate her. She argued that whatever she was now wanted to survive, and that she could be useful. The doctors were stunned when Sgt. Lowe spoke. He said there was no good reason to terminate her. She was a good soldier in the past and would be a good soldier now. Sgt. Lowe was not supposed to have any feelings on the matter at all. Dr. Smitherson told Sgt. Lowe said that it would be unethical to let Hunter remain when it was very clear in her file that she did not want to be reanimated. Sgt. Lowe quickly countered with the fact that not everything ethical is necessarily right or good.

Before the argument could continue, both Dr. Smitherson and Sgt. Lowe could hear Barny yelling for someone to come and get him. Dr. Smitherson ran out of the room and returned with him in hand. A conversation broke out between the four who could communicate with their minds.

"Why don't you just backdate her volunteer form?" asked Barny.

"It would be lying, and besides, her family must agree. Should I forget their signatures too?" said Dr. Smitherson.

"You don't have to," said Hunter. "My parents were killed two months ago in a suicide bomber attack in Nevada. There is no one left to sign."

Dr. Colman was getting upset as she tried to figure out what was going on. She knew a conversation was taking place and was aggravated that she was being left out of it by both Sgt. Lowe and Barny. They were demanding that Dr. Smitherson allow Specialist Hunter to remain reanimated and join the RAS.

Dr. Smitherson insisted Hunter did not want to be reanimated for religious reasons when she was alive. Now, he argued it was not really Hunter. Instead, she was a biological machine whose only concern was its survival. Spec. Hunter argued that it was not her wish but rather the wish of her parents, who were now dead. Suddenly, a loud, echoing voice yelled, "Talk out loud. I can't hear you."

Dr. Coleman had reached her wits' end and demanded to know what was going on. Dr. Smitherson apologized and promised that he would never again carry on a conversation telepathically if there was someone who should be in the conversation and who could not communicate that way. He brought Dr Colman up to speed, and she suggested things continue as if it had never been noticed in her file that she declined RAS service. Oversight was far less punishable than forging documents. She went on to remind Dr. Smitherson that, as long as she was not hungry, she performed like Sgt. Lowe. She would be very useful on the battlefield. Dr. Smitherson was about to give his approval when a loud noise once again interrupted the conversation. This time, it was the alert sirens. The air base was being attacked.

Lancegoroff committed nearly six thousand troops to taking the base and gathering information and intelligence, especially information about the RAS. The base was divided into several sections and was difficult to defend. The troops attacked in waves, but the waves were in very close succession. Soldiers from the base would be engaging a wave, and the next wave would charge forward, leapfrogging over the previous wave.

The timing was never the same, making it impossible to know when the next wave would move from cover and charge. Two or three would be lost each time a wave moved forward, but they kept moving.

As the main body slowly advanced, a Spetsnaz team was making its way to the RAS laboratory. They were using old diagrams of the base and building gathered by spies in the 1970s. The plan back then was to capture an alien spacecraft and an alien being. The plan was discovered, and the craft and beings were moved to a still-secret location. Now the plan was to capture Dr. Smitherson and as many members of his team as possible. Those who could not be captured were to be killed. Ichor was also a prized target. Even the smallest amount would be helpful to the communist cause.

With nearly every gun at the gates, the Spetsnaz moved uninhibitedly through most of the compound. They would easily shoot security forces, who were assigned the task of making sure no RAS became free, and not looking to make sure someone did not get to the RAS. Finally, the Spetsnaz was close enough to Sgt. Lowe, Hunter, and the two doctors realized that the shooting was inside and not part of the fighting outside. They ran to the lab as quickly as they could. Moose recognized the sounds of war and was running circles around the group, waiting for the command to attack. They entered the room and secured the door just as the Spetsnaz was coming into view.

Dr. Colman kept asking what to do. She was in a state of panic. Dr. Smitherson got her to calm down long enough to say in a voice that sounded and flowed like he was reading from a technical manual, "This problem requires a military solution. Military solutions are outside the scope of my expertise." Dr. Colman was dumbfounded. She wanted to hit him for such a ridiculous answer, and yet she knew the answer was correct. Sgt. Lowe shouted for the two doctors to lie flat on the floor.

He then ran to an autopsy table and pulled it loose from the floor, dragged it over to the pair, and stood it up on its side in front of them. Dr. Smitherson looked behind him for damage as bullets were now penetrating the door. He shouted, "Oh no," when he noticed a rack containing six vials of Ichor sitting on a lab table. They were going to be tested and placed in a large safe until it was needed to reanimate the division. He called for Sgt. Lowe both verbally and telepathically, but there was no answer. The noises were getting so loud, and bullets were bouncing everywhere, that Dr. Smitherson finally gave in to fear, put his hand over his ears, and lay down as close to Dr. Colman as he possibly could. She was in the same condition, and both were screaming at the tops of their lungs. Soon, the lights went out, the doors burst open, and flashlights began whipping through the room like an old searchlight at an airport during a storm.

The shouting stopped, and now footsteps from several different sources could be heard making their way towards the doctors.

"Get up and throw down your weapons," came a command in a clear Russian accent.

Now footsteps could be heard behind the doctors, and they knew they were no longer hidden. They stood up with their arms raised high.

"Well, well. Who do we have here? I think I have a good doctor," said the same voice that spoke before.

Dr. Smitherson wondered what happened to Sgt. Lowe, Hunter, and the dog. He got his answer when he looked towards the freezer where the RAS was stored and saw it slightly ajar, and he smiled as he wondered what Sgt. Lowe was up too. He then dropped his head in sadness when he saw a flashlight beam land on the vials of Ichor. The same voice dispatched a soldier to get the vials and two others to secure the doctors for transport.

As the soldiers moved towards their objectives, Moose charged from the freezer towards the soldier going for the Ichor. Gunfire erupted as the soldier increased his pace from a walk to an all-out sprint. He reached for the Ichor with one hand while maintaining fire on Moose with his other. Most of the rounds missed, but two struck Moose in his front right leg, shattering the bone. The soldier stopped firing, errantly believing he had incapacitated the dog. He walked up to Moose and pointed his 9 mm at his head and chuckled as he started to squeeze the trigger. Moose then rolled into the soldier's legs and knocked him to the floor. All the flashlight beams were now focused on the life-and-death struggle unfolding near a table. The soldier dropped the rack of vials, and all but one were now broken; the contents flowing onto the floor. Moose attacked first and was now violently biting and shaking the soldier's gun hand. He was desperately trying to pull his knife from its sheath. The soldier freed the knife and swung it around, shifting his body to give the blade more velocity. He could feel he was losing his grip on the pistol, so he just started repeatedly squeezing the trigger, expending all the ammunition in the hope that one would hit the dog. His knife hit the dog just behind the front right leg and penetrated up to the handle. He expected the animal to release as it ran to lick its wounds and die. Instead, he could feel the dog's lips quiver against the skin as the dog began to growl. It was biting down harder now as if the knife just made it angry. Soon, a loud cracking sound could be heard as bones in the hand and wrist were stressed to their limits and broke. The soldier shouted for help as the dog shook his arm until the hand separated from the arm. The soldier, now free, retrieved the unbroken vial and started running towards his comrades. Moose started to chase, but Sgt. Lowe telepathically called him back.

Sgt. Lowe crawled out of the cooler, captured the soldier's rifle, and was now lying in a prone position and using the night scope to select his first target. He had about half the pressure required to move the trigger backward applied, but he was waiting for something...a signal. Two soldiers were binding the hands of Dr. Smitherson and Dr. Coleman while a third was comparing them to photographs to make sure they had the right people. As this was being done, three soldiers screamed out simultaneously. Hunter and two other RAS troopers had grabbed their heads from behind and bit out large sections of their necks. Blood sprayed onto everyone close to the three men. Sgt Lowe counted eight men, and he began shooting them using the captured rifle. Dr. Coleman screamed as bone and brain fragments slammed into her face when Sgt Lowe terminated her captor. Within seconds, all the Spetsnaz were dead and accounted for except the one who lost his hand. He somehow stopped the bleeding and was long gone. He made his way to the gate where the battle continued and climbed into an unmarked car. He started the car and drove away. Not one of the Communist soldiers attacking the base tried to get inside. It was all just a cover for the Spetsnaz team to get in and get their targets. Every enemy soldier capable of moving withdrew from the fighting as soon as they received word that the Spetsnaz operation was completed. The Spetsnaz failed when it came to capturing people. Dr. Smitherson and his team were shaken, but none of them were injured or captured. They all knew they owed their freedom, if not their lives, to five zombies.

The four RAS troopers ate what they wanted, and the two recruited from the freezer were returned. Dr. Smitherson asked Sgt. Lowe, how he got them to help. Sgt. Lowe explained that he went into the freezer and telepathically asked if anyone could hear him. Those two responded and agreed to help if they got to eat.

Others responded, but only two were needed. Dr. Smitherson then asked if those two were as trustworthy as Sgt. Lowe, Sgt. Lowe said no because they were in it for survival only. He and Hunter were prepared to stop the two zombies after the soldiers were dead. They were surprised to see them settle for just one food source a piece. Dr. Smitherson concluded that having the actual death caused by the Ichor made a much more efficient and trustworthy RAS than injecting the Ichor after death. He thought about the times that a soldier who was mortally wounded on the battlefield and had volunteered for RAS service could not only have a physical impact on battle but could also have his mind and experience on the battlefield as well. Despite what just happened, Dr. Smitherson took out his notepad and pen and wrote down a note to discuss the morality and ethical questions of giving Ichor in the field to mortally wounded troops who volunteer.

Power was restored and, as soon as the lights came on, Gen. Lancer called for a meeting to finalize attack plans and to assign military units their specific mission and objectives. He asked Dr. Smitherson and Sgt. Lowe was to attend because he believed their knowledge of RAS abilities would be beneficial in planning an objective for that division. On his way to the meeting, one of Dr. Chi's assistants stopped Dr. Smitherson and told him that the broken vials were pieced back together, and it appeared that a vial was missing. He thanked the assistant and then started running terrible scenarios through his mind. He always said that what was in a vial was enough to reanimate one person, but such a small amount was used in Hunter's treatment that it seemed like a vial could reanimate 15 to 20 people. He thought of the possibility of zombies being unleashed on civilians, how horrible that would be. Then he wondered what would happen if the Ichor was reverse-engineered. What a detrimental effect that would have on society.

In the privacy of his thoughts, he compared it to a time when Iran did not have nuclear weapons and, despite all the evidence, some politicians thought they should be left alone. Now look at all the devastation because they were left alone. Some countries should never be allowed to possess large-scale weapons because of the country's leaders and their history of using them on innocent people. It was so simple to him, and he could not understand why others could not get it. Now, he was remorseful as he realized his carelessness may have just given a madman a very dangerous weapon.

As those attending the meeting were assembling, Dr. Smitherson's team was assigned a detail to start cleaning up the lab and taking care of the bodies. The dead Americans who volunteered for the RAS were taken to the lab, while the rest were stored in a makeshift morgue. Enemy combatants were stacked in a freezer until it was full. The rest were buried in a mass grave. Neither the Muslims nor the Communists reclaimed their dead. Rumors abound as to what the bodies in the freezer were going to be used for. Food was the best guess, but experimentation ran a close second. It was both. The team was trying to find a way to produce food for RAS troops that did not have to come from a human. They also wanted to know how to cool the body to near-freezing stored energy in both a dead body and a reanimated person. There was no explanation for this, as electricity could be conducted through ice.

Medical teams were sent out to treat the civilians wounded in the fighting. Many of them had never seen war except on TV. The bullets and shrapnel killed and wounded indiscriminately. The very young, the very old, and every age group in between lay dead or wounded in the street. Many of them just wanted to see what a battle looked like. They had watched it on TV safely for so long that the mechanism in their head that decided this was dangerous was no longer working. It was one of the risks a person took when choosing to live so close to a military installation. Sooner or later, even the most secure base was going to be attacked.

Back at the conference room converted to a war room, Gen. Lancer laid out his plans.

"Sgt. Lowe's idea of using live soldiers disguised as RAS on the Muslims was a go. It was hoped that this tactic would spare the human shields but still result in the death of leadership and enemy combatants. The 82nd and Rangers are being briefed and trained as I speak for this important mission. This attack, combined with massive bombings of enemy positions in California, The Vatican, and the Middle East should leave the Muslim Brotherhood, and all its allies crippled and send them home with their tails between their legs. We can end a war that has lasted over fifteen years today. I have the President's approval to suspend all rules of engagement. If it looks like the enemy, kill him or her. Do not let them shoot first, and every round they shoot at you, shoot seven back at them. In short, and as it used to be in the old days, kill them all and let God sort them out.

Special Ops has already left to take care of the nuclear threat in New York. After killing the troops guarding it and disarming it, they will meet up with civilian resistance fighters. Those fighters will lead them to small targets hiding around the city. They will engage and destroy as many as possible. Special Ops will have a squadron of Army Apache Helicopters and two companies of Marines at their disposal and should be able to seriously disrupt the enemy's ability to coordinate any defensive maneuvers. The main attack on the communist will be directed towards their headquarters. The plan is to bomb headquarters first and then send in RAS 101st and the 1st Infantry Division. The Big Red One is already in New York and can reach Headquarters in six hours. Lancegoroff has brought nearly all his troops to New York in preparation for a showdown.

China and Russia have committed troops, so this will not be a walk in the park. Both the 1st and 101st soldiers will be wearing what we are calling zombie-proof uniforms. No other units should go near the battle at headquarters until you are called to do so. Squad leaders will have a safety device that will kill RAS soldiers should the suits not work or if something else goes wrong.

There will not be any armor support except for the attack on the headquarters. The fuel shortage is grounding armor and is starting to ground aircraft. It would be pitiful if we lost this war because we were too stupid to refine our oil, but that is a possibility. Written instructions will be provided to each division commander. This may be our last hope, so lead at your best level and tell your troops to fight at theirs. Start time is 2200 hours."

He was about to dismiss the meeting when three men in suits entered the room and turned on a large monitor behind Gen. Lancer. "Sir, you will want to see this," said one of the men as he turned on Fox News Channel. Sen. Mikevell was speaking from the Capitol steps, waving a stack of papers over his head.

"We have a decision that says, in short, the court finds that initial attempts to vacate the SEFWMS were executed illegally, and no other action should have been taken until the court reviewed the case. That means all the money that has been taken from the SEFWMS and spent on war must be returned immediately.

It is a sad day when the President of the United States violates our country's laws and spends money designated to bring the poor in this country to a better economic status, and then uses the money to kill others."

The reporter interrupted, "Are you not concerned at all, Senator, that the people we are fighting kill Americans every day, or with the fact that they have captured California, your state, and destroyed another?"

"Well, the most important part of your question is the fact that they are from my state. I represent them as well as other Californians. I do not like what has happened, but they are still my constituents. They are fighting to live a life that feels right and just, much the same way that George Washington and his followers did."

"I'm sorry to interrupt the Senator, but I want it clear for the record. You are comparing the likes of Osama Bin Laden and King Abdulla to George Washington?"

"Yes, but only in the sense that they are willing to fight and die to make things better."

"So, you're saying the way women and girls are treated, as an example, would be better for our nation?"

"I'm saying they believe it would be better, and I do not think it is right to punish them for that."

The Senator continued.

"As far as I can tell, any further use of the RAS is illegal because the money spent on the program was done so illegally, and any more military action against the citizens of California is illegal by nature of the fact that they are citizens.

I call for an immediate end to conflict so we can bring California back into the fold peaceably and while respecting the beliefs of our Muslim citizens."

The Senator then left the speaking area and disappeared down a hallway.

Simultaneously, the reporter stated, "There you have it, folks, a United States Senator looking out for the wants and needs of an enemy who has killed more Americans than any other. We report, you decide."

General Lancer's face turned as red as a beet. "This political correctness nonsense is going to kill us all," he said under his breath as he reached for a single phone sitting on a table by itself.

"Mr. President. What is going on, and what do you want me to do?"

Gen. Lancer said, "Yes, Sir" a few times, followed by a "Will do, Sir," and then he hung up the phone.

"Listen up. The Speaker of the House is working with the President. He has called for a special session of Congress to discuss this mess and has given us 24 hours to get it done and get back. The money is going to dry up after that until the court rules. It has come down to this. We win tonight or lose over money and politics. There may even be some who are charged with a crime when this is all over, so if you don't want that on your record, leave now." He paused for a second, but no one moved. "Alright then, let's get the planes filled, magazines loaded, bombs falling, and the enemy dying."

Crews around the world worked vigorously to get things ready. Those loading the planes with RAS moved over six thousand dead brothers and sisters onto planes. These were not refitted planes either. The IV system hung from overhead and often fell as they were being attached to the needles in soldiers. Plywood was used as short walls to help prevent the RAS from tangling up in the seating and cargo systems that had not been removed. Many of the planes only had enough fuel to reach the target, drop their bombs or troops, and be ditched in the ocean. Crew members would have to run to an open door and parachute out before the plane crashed. Many people were being asked to do things they had never done before and things that had a low chance of survival. It appeared the odds had never been so highly stacked against the United States Military or so much riding on one operation.

The remaining minutes passed, and it came time for the first planes to take off. As they climbed to altitude, they were detected by enemy radar. As soon as the first plane was detected, Sen. Mikevell appeared on the monitors. "General, I am ordering you to stop this madness now. You will be arrested if you go through with this."

Gen. Lancer did not respond. The planes and troops carried on towards their targets.

After about three minutes of ordering Gen. Lancer to stand down, three Muslim fighters entered the room. One was wielding a large kitchen knife. Two of them grabbed the Senator and put him on his knees while holding him in place by his arms. They chanted lightly in Arabic as the third one took up a position behind the Senator, who was bowed between his legs. The Senator's face turned white as he realized the error of his political position. The man behind the senator ruffled the Senator's hair with his empty hand and then grabbed it and pulled the hair back, forcing the Senator's head back. Senator Mikevell began to scream as he realized what was happening to him.

His eyes peered into the camera, pleading for whoever was watching to save him. The man with a knife put the knife between his teeth, pulled a handkerchief from his pocket, and stuffed it into the Senator's mouth. He kept pushing until the screams were muffled, and then he took the knife back in his right hand and pressed the sharp edge against the Senator's neck, slowly pressing into the skin until a small trickle of blood began to flow.

"You have one minute to issue a recall to the commanders, or your Senator dies. "

The other two men started counting backward from 60 in Arabic as Senator Mikevell's face flushed, and he closed his eyes in anticipation of his death. Sixty seconds later, the man with a knife began forcing it into the Senator's neck. He fought and screamed as loudly as he could over the handkerchief, but he was dead in just a few seconds. The man continued to cut until his head was completely severed. The three men stopped in a puddle of blood as they dropped the body, and all three grabbed the head and held it aloft. One cried out, "God is great," and then they all started chanting in Arabic as the camera slowly faded to black.

"Why do people have to learn the hard way?" Gen. Lancer said angrily. He did not like Senator Mikevell at all, but he was an American and did not deserve to die that way. He wondered what the Senator thought when he realized the enemy was not his friend and was going to kill him.

"Enough of that," he thought, "I have a war to fight."

Chapter 11
One Last Hoorah

The continuous whining of aircraft engines could be heard all over New York. There were so many planes in the air that the hardest communist soldiers were trying to fortify their positions. The first gunshots, however, echoed throughout the subway and sewer systems. It did not last long as the SEAL team quickly neutralized those guarding or working on the nuclear bomb. Most never knew they were killed; they just fell lifelessly to the ground. The explosive expert on the SEAL team quickly recognized the fuse type and disarmed it. She also noticed that it was very well booby trapped and said it would have to be removed later. A security team was called in to guard the bomb, and the SEAL team continued to its link-up point with the resistance fighters. There were many more fighters than expected, increasing the chances of success. Most had experience and were well armed, but some came with swords designed for decoration or a garden tool of some kind. They did not plan on surviving, but decided they could at least draw enemy fire while a better-armed and equipped fighter moved closer to the enemy. This was a tactic being used by the resistance for some time now, but the Government never knew.

It was not long before the helicopter's gunfire and explosions filled the streets. The resistance produced some very accurate maps containing enemy compounds and strongholds.

The force grew stronger as it progressed, even picking up a company of Marines who were long ago thought lost. They were assisting the resistance, distributing captured ammunition and weapons, and just waiting for a chance to get back into the fight full swing. The SEAL team leader used the maps made by the resistance to call for missile strikes on larger targets.

The maps were so accurate and updated that the need for building-to-building searches was not necessary, saving countless hours of dangerous and tedious work. Though most of New York City lay in rubble, there were still thousands of civilians hiding in the buildings that were left. Clearing them all would have cost lives and time.

As the SEAL team was moving between FSS, what the military named former skyscrapers, reduced to mostly rubble but still being occupied, they came under sniper fire. The sniper's location was quickly identified. He was shooting from midway up a 30-story building through a broken-out window. He had a spotter with him, and his shots were getting closer to team members. The team leader got on the radio and called for air support, and within seconds, an AH-64 was heading towards the building. The helicopter was coming down at the target and headed straight for it when the sniper laid down his rifle and picked up an RPG. He aimed and fired as his spotter ran for cover from the back blast. The grenade traveled up towards the helicopter, leaving the familiar trail of white smoke as it progressed from the shooter to the target, missing the main body but striking the tail rotor. The pilot fought to maintain control as he continued toward his target. The sniper had now retrieved his rifle and was firing at the helicopter's canopy. The damage caused by the grenade and the bullets was increasing by the second. The pedal steering was virtually useless, and the collective no longer controlled the speed of the helicopter. It was all the pilot could do to fly toward the target. He thought of using the stick to break away from the course and crash into the street, but his sense of duty would not allow him to do that. The communication system was now out, so the pilot kicked the gunner's seat to get his attention. The gunner held up his right thumb, knowing that the pilot was telling him they were going to crash into the target.

The gunner turned his eyes towards the target and depressed the button to fire the 30 mm gun. Utter disappointment filled his head as the weapon did not fire. He quickly switched to the last two HellFire missiles, locked on, and fired. A fireball shot from the window as the two missiles impacted a split second before the helicopter, and the helicopter disappeared in the fireball. The pilot pulled back on the stick out of instinct, and the helicopter leveled just before slamming into the ground. The rotor was mostly gone, having been eaten away by the walls of the building, but the bird was otherwise intact as it slid out of control across the floor. It finally hit a support column, breaking off the tail section, and came to a grinding stop. The pilot reached up to turn off the engine when a cold rush of realization overtook him. He should not be able to turn off the engines. He should be dead. He kicked the gunner's seat again as the engine ground to a halt. The gunner once again held up his thumb. The canopy would barely move and needed to be pushed open. They climbed out, drew their 9MMs, and moved towards the sniper's location to make sure he was dead. All that could be found was a right arm still attached to the badly damaged SVD rifle. There were no traces of the spotter. The search for the sniper took them near the window where they could be seen by a large audience of SEALs, Marines, and resistance fighters.

The two did not know they were watching and embraced and danced a little in much the same way a pitcher and catcher would do after a perfect game. They quickly regained their composure, and the pilot turned on his emergency radio, pressed the talk button, and professionally said, "Sniper threat neutralized." The SEAL team moved towards the building to help the Apache crew, but they were out of the building when the team arrived at the base. The pilot and gunner were debating how they survived. The three most accepted theories were that the pilot leveled the bird, and the energy stored in the mass of metal was transferred into a slide rather than destroying the aircraft. The blast caused by the HellFire missiles interacted with the disc created by the rotors and slowed the craft down, or God and country were not done with them yet.

Whatever the reason behind their survival, word of the incident spread quickly, and it was seen as a turning point for the battle for New York. Everyone believed they were invincible and that they would win. Not invincible as individuals, but invincible as a cause. As more enemy combatants died, more freedom fighters found weapons, and the cause grew in numbers, firepower, and spirit. Soon, enemy soldiers were surrendering in large numbers to people armed only with a shovel or a kitchen knife. They were amazed at how their wounds were treated, and they were offered clean water and food. This kind of compassion was not what they were told they would receive if they surrendered to the Americans. Emotions soon became uncontainable as victory was no longer hoped for but realized. American flags were attached to handles and waved in the streets as New York was once again an American-held city, and in only six hours and thirty-seven minutes from start to finish.

As New York was being taken, other battles raged around the world. Small units to whole divisions attacked enemy positions and bases. Fighting was intense and death gripped all three armies, but the United States, by far, was winning. Navy ships launched missiles to soften larger or dug-in targets, and the Air Force bombed enemy positions that were attacked by troops into piles of bricks, splintered lumber, and chunks of concrete. Muslim units were hardest to overtake because there were more of them. Communist positions collapsed quickly because Lancegoroff had been massing troops in New York for some time. He believed his plan would work because he thought most of his troops in other countries would only be fighting Muslims. He still did not know about the large-scale defection of his troops to the Muslim side.

Lacegoroff was dead set on massing troops on the East Coast of America and then moving across the country, taking it all. Leadership had gone to his head, but he never lost sight of his belief that Americans were lazy and always demanded more, no matter how much they had. He thought North America's resources would be better used by him and the communist way of thinking.

As more and more reports of success poured into Gen. In Lancer's war room, more and more commanders became confident that the war would soon be mostly over. Some of them began making plans for mop-up operations. Gen. Lancer decided to reserve his judgment until he heard from the 82nd commander and General Lincoln. Dr. Smitherson and his team were on edge, expecting news of the RAS every time a phone would ring or a radio would squelch. Tensions were still high in the room, but were relaxing with each report of success. Gen. Lancer reported that the body of Senator Mikevell was found in a public restroom in the Capitol Building. The murder was believed to have taken place there because of the amount of blood present and the absence of drag marks. This would mean that the murderers are employees of the Capital because no one else got in but employees. The Visitor's Center has been closed for years. He wonders how far into the system they penetrated and who else had such easy access to the Capitol.

The 82nd Airborne and the Rangers were nearing their drop zone. It had been a long flight for the men and women as they thought the entire time about the task that lay ahead and what could go wrong. Young officers tried to reassure their soldiers, but the look of worry was showing on their faces as well. The jump itself was dangerous. To make the ruse work, they had to be "dumped" from the plane and land as awkwardly as possible.

Next, they had to appear to kill women and children with the enemy watching, hoping intelligence was right and they would not get shot in the process. The makeup was uncomfortable, and many struggled against the urge to remove it.

The commanders received word on the planes that the building was surrounded by women and children and that all military combatants were inside the building. Thumbs up moved through the planes, letting everyone know that the mission was a go. Soon afterwards, yellow lights came on indicating that the drop zone was near. Everyone on the planes stood up and prepared for what was sure to be a wild ride. They braced against each other as the planes began to climb and the ramp opened above the drop zone, while they climbed rapidly. One by one, the planes emptied their cargo of (living) living dead soldiers. There was no resistance of any kind. No AA or small arms fire. Not even a radar lock to let the pilot know they were watching. From the air, soldiers could see women trying to be brave and stand around the building, clutching their children tighter as the soldiers neared the ground. Small explosions could be heard as the paratroopers hit the ground. Parachutes were carried away by the wind. Those who landed inside the fence wandered until all those who landed outside the fence slowly made their way into the perimeter.

Once everyone was in place, one of the captains moaned in an unnatural voice and then started sniffing the air. Soon, 1100 fake zombies surrounded the large building and were sniffing the air and moaning. The children were screaming in terror as their mothers held them tightly and prayed. The soldiers moved in on the civilians, turned them away from the windows, and held rags with chloroform over their mouths and noses until they became unconscious.

They then carefully laid them on the ground, poured animal blood on them, and acted like they were eating them. They looked back occasionally so they could be seen with bloody faces. Some of the soldiers were so upset by the children's screams that they, both men and women, cried themselves as they forced the children to go to sleep.

After just a few minutes, all the women and children lay on the ground unconscious. The soldiers walked around a little and then started one by sitting down and closing their eyes. Several soldiers took up positions behind cover so they could serve as snipers without immediately being detected by the enemy. They would also be able to return fire quickly should the enemy run out of the building, shooting. With all eight exits and the one-story building covered, it was time to wait for the enemy.

Inside the building, 400 soldiers line up in single-file lines, 25 to a door. Their instructions were to form a firing line in front of the RAS and, when the command was given, fire simultaneously into the heads of the RAS. Every soldier waiting in line was justifiably terrified at what they were about to do. Some saw the video of the last attack in California, others heard rumors, but they received the news; they all could die a most painful death if their plan failed.

As they pondered their fate and tried to overcome their fear with faith and humor, a blast from a small plastic horn mounted on a can of compressed air spread the news that it was time to go outside. All eight doors opened, and 25 soldiers passed through each. They walked along the exterior walls of the building just in front of the RAS and quietly waited for the signal to open fire. The stand-in zombies were waiting for their signal to pull out their weapons and open up on the Muslims. The tension could be felt in the air.

The Muslims had no real faith in their plan, believing that as soon as the first shot was fired. The RAS would rise, kill them, and eat them. The Americans wondered how they were going to draw their weapons and get off the first shot when they had no idea when the enemy was going to start shooting. The intensity of the standoff increased dramatically in the 15 seconds it took for a Ranger positioned behind a truck to notice a man climbing through an opening on the roof carrying the air horn. The Ranger quickly surmised that the horn was the signal for the Muslims to start shooting. He quickly drew a bead on the man and shot him dead. Gunfire was the signal for the Americans to start shooting. They all reached inside their unbuttoned uniform tops, pulled out an M-16, and blasted the enemy. Muslims were completely caught off guard, many still waiting on the horn as bullets ripped through them. They were all killed with only minor injuries to American troops.

The sound of gunfire awoke some of the women and children. Muslim leaders were looking out of windows, thinking the shooting was their soldiers, and it was safe because the shooting stopped. Instead, they saw some of the women and children setting up and American soldiers wiping makeup from their faces. Their hearts sank, and lumps formed in their throats as they realized they had been duped. Shouting could be heard as those still inside began barricading the doors with furniture. One man ran out of the door, and he was killed by a sniper before his second foot hit the ground.

Without a word being spoken, several Americans took up positions around the building to watch for and kill anyone who tried to leave. Others started carrying the women and children about 100 yards away and treating any injuries they may have had.

Still others began placing explosive charges at five-meter intervals along the walls, with extra ones at the corners. There was some shooting through the windows, but it was inaccurate and was always quickly suppressed with a high volume of friendly fire. Once all the charges were set, most of the Americans moved to the location with the civilians and began waking them with smelling sauce and helping to treat their wounds. The remaining Americans began breaking out the remaining windows and tossing in incendiary grenades. One Ranger approached a window that already had most of the glass broken out. A white piece of cloth came out of the window, attached to a flagpole. The Ranger took out his cigarette lighter, lit the cloth on fire, and pulled the pin on his grenade, chucking it through the window. Smoke was billowing from the windows now, and screaming could be heard from inside. Several men pushed furniture away from the doors and tried to run out in groups, thinking one or two might make it. They were wrong. The snipers were ruthless and did not miss.

As the last American reached the rally point, a soldier held up a remote control and depressed the button. The ground shook as the sound of the explosion filled the air. Soldiers wrapped themselves around the children, trying, as much as possible, to let them know they were going to be alright. The thick concrete walls of the building buckled, and the ceiling collapsed on those who were still inside and suffocated the fire. When the smoke and dust cleared, a squad returned to the building, looking for anyone who may have escaped. There was no one outside the pile of rubble, but a few screams could be heard coming from beneath it. The soldiers stood in place until the screams subsided, then they returned to the group.

Soldiers then reached into their cargo pockets and pulled out MREs and bottles of water. They gave the supplies to the women along with Tylenol and infant Tylenol to help with the pain they might be suffering. An interpreter explained to them that they were going to be alright and that they could eat the food and drink the water. They were also told that more Americans would be coming and that they could wait for them for help or leave on their own.

The soldiers gathered in their squads, shared some high-fives, and quietly started walking towards the extraction points. Engineers and Intelligence units were already en route to sift through the rubble and identify those who were killed. There was also hope that some documents survived the fire and could be used in future operations, if there was a need for any. The mission commander delivered the news of success to Gen. Lancer using an unsecured cell phone. This was done intentionally to allow what was left of the Muslim military to intercept the message and realize that their leadership was dead, and the rest would be soon. Rockets, missiles, and bombs pounded other Muslim positions throughout California and around the world. Soon, Muslim soldiers were laying down their weapons and surrendering to anyone they met, much the same way Iraqis surrendered during Desert Storm. Victory could be sensed now from New York to San Francisco. The more news of winning that came in, the more the will to win grew. The only major news needed now was the results of the attack on Lancegoroff's headquarters.

Lancegoroff's headquarters were formidable. It was a piece of ground just outside New York City, three miles long and two miles wide. It housed two Divisions and a Spetsnaz unit. Tanks were dug in around the perimeter so deep that only the turret could be seen.

None of the tanks had enough fuel to be moved, but they all had enough fuel to power up and use the main gun. Just past the tanks were dug-in soldiers. Their foxholes were very well constructed and fortified. A machine gun was in every third foxhole, and each position had a well-placed claymore mine thirty yards in front of it. The entire network of tanks and foxholes was wired for communication, and the wires were buried to prevent them from being cut by the enemy. Getting through the defenses will be a challenge for any unit.

The troop barracks and small buildings had different color shingles on the roofs and mismatched paint on the walls because of repairs to the damage caused by bombing. There were two foundations left from buildings that were destroyed. Most of the vegetation was dead or dying. Burnt and broken trees dotted the landscape. There was an occasional tree that somehow was missed by the explosions, bullets, and fires. What little green on the ground was more crabgrass than the tall fescue that was there before the war? Most of the ground was cracked dirt and rocks. There was a small runway littered with car-sized potholes. A small Cessna sat at one end, pointing down the runway as if it were trying to escape, but never got the chance.

The main structure was monstrous compared to the other buildings around it. The building was seven stories tall above ground and had an unknown number of levels underground. The façade of the building was taller than the building itself, and it included corner towers that were of no use other than decoration. The entire building was constructed of solid concrete blocks laid around a steel frame. There was an arch over the main entrance, and green bushes stretched from either side of the arch to the corner of the building. There was green grass for twenty meters in front of the building. The grass was well watered and maintained.

It was clear that the building had been painted over, maybe several times, but the grey paint still gave way to high burn marks going up the sides of the building. The marks resembled a giant chart showing the ups and downs of a company's sales. The grey color of the walls contrasted with the neatly maintained lawn and manicured shrubbery. The building could easily remind someone of a wicked witch's castle. Security around the building was not very prominent. That task was entrusted more to the tanks and the men in the foxholes.

Just off the grass and at each corner of the building was an anti-aircraft gun. They were constantly manned, and the crews trained every day. On the roof of the building, there were always two snipers and a stinger missile crew. Inside was a company of infantry and the Spetsnaz team. They were the last stand to defend Lancegoroff. Every soldier was mentally ready for what they knew was coming. They knew the battle would be hard and many of them would die.

The American plan called for 3000 armed RAS troops to be dropped first. This would allow for the weapons systems to kill as many enemy soldiers as possible without risking American deaths by friendly fire. So many people had never been dropped in such a small area before. Twenty-eight planes that were not properly fitted for the job were airborne and had a hundred or more newly reanimated soldiers in the back, kicking and moaning, trying to understand what was happening to them. As with the first mission, some were curious about their situation. Some looked at the outdated Battle Dress Uniform and wondered why they were wearing it. Others stared at the weapon attached to their arm and tried to figure out how it worked. Still others were just hungry and shifting around as best they could to find an energy source. As the planes approached ten miles out, a naval ship launched four Tomahawk missiles programmed with the coordinates of the anti-aircraft guns. Guns at those positions had been destroyed before, but were quickly replaced.

The goal was to give the pilots enough time to dump their cargo while new guns were being set up to replace the guns destroyed by the missiles. The missiles hit their targets just as the planes were close enough to see the fireballs. They started quickly dropping in altitude so they could climb above the target area and deploy the RAS. There was no anti-aircraft fire, but Lancegoroff did have a battle plan for this scenario. He ordered his tank crews to fire at any aircraft they felt they could hit. The planes were flying low enough that it might be possible to hit a few of them. Pilots could see muzzle flashes as the planes approached a mile out and got ready to start climbing.

A round from a tank found the right wing of one of the planes. An explosion filled the now dark sky with temporary light as the plane slammed helplessly into the ground. The crew died on impact, and most of the RAS in the back walked around burning painlessly until the intense heat melted their brain stem. Finding a food source was still paramount as their bodies burned to small piles of ash and bone. Seven of the RAS somehow escaped the fire and walked about in small steps with their nostrils flaring in search of the smell of human-generated electricity. One of them saw a muzzle flash and somehow realized that a food source could do that. He started walking towards the tanks, and the other six saw him and followed. As they drew near a tank, they detected the energy of a human and saw a tank commander standing near his tank, ordering some of the crew to get out and help swab the gun tube.

The planes were all past them now, and they needed to get ready for the tank battle they knew was coming. The gunner stayed inside, scanning his sector with the thermal imaging sight. All he could see was his crew. The loader and driver jumped to the ground just as the RAS came into view of the commander. He pulled out his pistol and emptied his magazine, but the bullets never slowed down the RAS. They rushed in and killed the crew, ripping the bodies apart and eating strips of flesh the way some people eat a large piece of beef jerky. One of them realized the strongest source of energy was coming from the head. Soon after, all three heads were broken open, and the brains were gone.

The gunner at first tried to locate the attackers through the sight, but, in panic, never turned off the thermal imaging sight. He never saw the enemy, so he did not know what was happening. He slowly raised his head above the gunner's hatch and peered out. He froze in horror as he saw his crew being eaten by the RAS. He slowly lowered himself back into his seat and picked up the phone to warn the others when something told him to look up. When he did, he saw a RAS staring down at him with blood and small pieces of flesh falling from the zombie's face onto his. He quickly grabbed his pistol and put the barrel under his chin.

"I gave my best" were his last words spoken in Russian just before he pulled the trigger.

Six of the RAS sat down and closed their eyes for rest, but the one who followed the muzzle flash walked around the tank, fingers slightly touching it as he circled it once completely. He had flashes of riding in one. The sounds of it being fired and the fear of it being fired upon were just short of overwhelming. He remembered how it all worked. It was a captured M-1 Abrams.

He remembered volunteering for an experiment involving continued service even after death, but he thought that meant they would use his organs for transplant or something. He pieced together what was going on and knew he was in a battle. He figured that the planes flew past the tanks, so the enemy must be behind him. He reached into the gunner's hatch and pulled the dead man out, and rolled him off the tank. Using the hand crank, he traversed the turret 180 degrees and then peered through the thermal sight. He could see the cooling silhouettes of other turrets and gun tubes and the heat signatures of soldiers in their foxholes. He put an armor-piercing round in the breech and prepared it for firing. He planned to destroy the tank to his right. He jumped down and reloaded, destroyed the tank to his left, and then took out as many of the men in foxholes as he could before one of the other tanks got him. He locked in on the tank as low on the turret as he could and fired.

He saw through the sight that the turret rose and moved over, rendering that tank useless and the crew dead. He quickly removed the shell from the breech and loaded a new round. He jumped back into the seat, traversing the turret as he peered through the sight.

"Target identified," he heard himself say as he locked on and fired with the same results. He then scanned for heat signatures with his fingers ready to fire the coaxial machine guns. He fired on each partial silhouette, losing track of how many he killed. He continued until the ammunition was gone. He decided to get another can of 7.62 mm when he heard the deafening sound of an armor-piercing round impacting on the outside of the turret. It was followed by a second and then a third. Now he was pinned under twisted metal, and the heat was intense from the metal burning. He struggled for a minute but then gave up and waited for death a second time. In the light of the fire, he noticed a strange unit identification patch on his left shoulder.

He grabbed it and pulled it closer to his face. He saw the word "Second time around" and smiled. He then ripped off the patch and reached for the one on his right shoulder. The unit patch for the 11th Armored Cavalry Regiment. He ripped it off and kissed them both. He was glad that he would get to die twice for what he believed so strongly in. He was soon consumed by the fire. He did not know it, but he just opened a breach in the defense for the approaching 1st Infantry Division. The other six RAS moved away from the heat and continued walking towards the headquarters, unseen with the naked eye and undetectable with heat-sensing scopes.

At the same time, the short tank battle was taking place; planes were dropping RAS soldiers into the compound. Many were dropping their troops and then heading out to sea to crash into the ocean, while others were heading back to base. The dropped RAS was performing well. Parachutes were being released, and the metal rods in the boots were preventing broken legs. The weapons systems were killing enemy soldiers, but they were also locking in on fires and emptying the magazines on them. The RAS were killing and eating more soldiers than expected. It was thought that the weapons would do most of the killing. The battle resembled something more of a horror movie than a war zone. Enemy soldiers were going through ammunition very quickly because most of their shots could not stop the RAS. Bullets were pretty much useless. It took grenades or mortar fire to be successful.

The second wave of the attack was much larger and included both living and undead soldiers. Planes were dropping another 3000 RAS and Black Hawk helicopters were inbound with the 101st rested and ready to fight. One Black Hawk had three parachutists, Sgt. Lowe, Specialist Hunter and Moose. Sgt. Lowe and Hunter paced for two hours waiting to leave in an effort to make themselves hungry before the battle. The helicopters with the 101st were flying in a holding pattern waiting on all the planes to drop the RAS.

The RAS was going inside the perimeter while the 101st were heading towards the outside of the perimeter. They were to meet inside and make entry into the main building. Sgt. Lowe and Hunter were to parachute from the Black Hawk and attempt to control them using telepathic communication. The enemy divisions were falling in numbers rapidly. Most were dying, but many were just deserting. Fighting an enemy that couldn't be killed was not something they bargained for.

Sgt. Lowe, Hunter, and Moose were the last RAS on the ground. They immediately tracked down two enemy soldiers to kill and consume so they could gain better control and focus on the mission. Their prey did not even put up resistance because they were out of ammunition, and they had already seen what happens when you surrender to RAS troops; death comes quicker. After eating, Sgt Lowe determined that the interior of the perimeter had been won. RAS soldiers were finding places to sit down and rest. Sgt Lowe decided it would be good to attack the perimeter from the inside, but he needed to get the RAS back into the fight. By this time, 101st soldiers were attacking the tanks with AT-4s and trying to get to the foxholes. The 1st Infantry Division arrived right on time and added to the tank-killing abilities of the Americans. The problem was that the enemy tanks were so well-dug in and the men in the foxholes were well-disciplined, letting the tanks fight a tank battle. Lancegoroff did not pull them in to help inside the perimeter, but he certainly would if his headquarters building was about to be entered. The building could not be taken until the tanks and troops on the perimeter were neutralized.

Sgt. Lowe and Hunter decided to take what RAS would respond to a mission to attack the foxhole positions from inside the perimeter. Hunter would lead them while Sgt. Lowe worked on a way to enter the building. Three hundred RAS soldiers answered Hunter and Sgt. Lowe.

Some did not understand what they were doing, but knew they needed to answer. Most of the responding RAS understood completely and began gathering rifles and grenades. Hunter led them off into what was quickly becoming a beautiful sunrise, dampened by the smell of burning steel and flesh and the screams of the wounded and dying. Black columns of smoke rose through the air from burning tanks and buildings. Hunter and her team made their way to the perimeter. They moved from foxhole to foxhole, rolling in grenades and then shooting whoever tried to climb out. There were many more RAS than were needed, but still they quietly followed each other, watching and waiting for their turn to kill the enemy. When a RAS would feel even the slightest urges of hunger, he or she would stop moving with the group and eat.

Sgt. Lowe decided to use explosives to try to open the door to the compound; he searched several bodies until he found a pack with C-4 explosives and a few blasting caps. A sniper on the roof continuously shot at Sgt. Lowe as he looked for the explosives. The sniper scored several kill shots if he was shooting a living soldier, but the bullets passed right through Sgt. Lowe with nothing more than the sound of the round impacting flesh and sometimes bone. Three rounds hit him in the head, but the combination of a Kevlar helmet and a titanium plate proved too much for copper and lead.

Sgt. Lowe moved to the door of the building without any regard for tactics. He grabbed the bag of explosives, stood straight up, and walked at a quick step towards the door as the sniper's bullets continued to fill him with holes. The sniper made a bold decision as Sgt. Lowe disappeared under the façade of the building. He tossed a rope down and quickly donned rappelling gear. Sgt. Lowe was unaware of the rope, focused on having the door and as much wall as possible gone when the 101st got there.

The sniper descended the short distance quickly, and the sound of his boots hitting the concrete caught Sgt. Lowe's attention. He stopped moving just as he pushed the blasting cap into the explosive wrapped around the doorknob. He slowly turned his head towards the sniper, who was now looking at Sgt Lowe through the sight apertures on his pistol. Sgt. Lowe slowly stood completely upright, knowing he was finished but wondering how he could detonate the explosives before a bullet hit his brain stem. He knew the sniper was good, and he knew he would not miss. Just then, and much to Sgt. Lowes' amazement; a look of bewilderment quickly came over the sniper as though something horrific had just happened. In a second, he lowered his pistol. He rolled forward until his forehead met his lower chest, exposing the inside of his neck. The vertebrae were no longer intact, crushed, and separated. Arteries were exposed, but no blood flowed out because the heart was no longer receiving the message to pump blood. Only the skin across the throat and a few stringy pieces of flesh kept the sniper's head from falling to the ground. It all seemed to be in slow motion to Sgt. Lowe. The sniper was simultaneously collapsing to the ground as his falling head revealed the trauma to his neck. Standing behind the sniper was Moose. Her tail was wagging, and her tongue was hanging out. She reminded Sgt. Lowe of a dog, waiting for his boy to throw the stick. Sgt. Lowe smiled and thanked Moose out loud. He walked over and patted her head before finishing with the explosives. He walked backwards, unrolling electrical wire as he went. He found cover behind the left corner of the building and called Moose to him as he attached a firing device to the spool of wire.

He squeezed the handle without warning or any countdown. Several resting RAS squirmed a little at the deafening sound of the explosion, but then returned to a peaceful rest mode.

Sgt Lowe lowered his head as debris zoomed past him. When he did so, he noticed all the little holes in his clothing and began to wonder just how serviceable he was right now. He wiggled his fingers and waved his arms twice in short motions and said out loud, "I'm good." Moose rolled his head slightly right and looked at Sgt. Lowe with his ears pointing up as if to say, "What are you talking about?"

Once the debris stopped falling, Sgt. Lowe peered around the corner to see the size of the opening he had created. What he saw distressed him, and he struggled to control the overwhelming urge to panic. A condition he couldn't remember from being alive. He pulled his head back behind the corner, then peeked a second time to make sure he saw what he thought he saw. Unfortunately, he was right. Six Spetsnaz emerged from the now missing door with .50 caliber rifles. They wore very heavy body armor that extended past the neck. It was as if they had prepared and trained for this opportunity. They separated, and one by one, they shot the resting RAS in the head. The protective headgear was no match, as on many occasions, the RAS's head would explode. They were quick and methodical in their work. They couldn't kill all the RAS, but they were most definitely going to kill as many as they could. Sgt. Lose watched in disbelief as he screamed as loud as he could telepathically, but the RAS would not move. Sgt. Lowe asked Moose to move across the front of the building, and then they would attack from both flanks. He thought that maybe they could at least slow down the slaughter of the RAS until help arrived or they were both killed again.

Sgt. Lowe looked out over the dead until he spotted an AK-47. His first stop on the attack would be to secure that rifle. He waited to hear that Moose was ready, anticipating the joy of picking up the rifle and getting some payback.

Moose instead telepathically sent words of distress. Sgt Lowe caught sight of her running from the building with six large pit bulls in pursuit. They were beautiful dogs. Two were black and white, and the remaining three were deep brown. They were well cared for and well-trained. They also had five-inch-wide leather collars around their necks. Metal spikes at least four inches long protruded from the collar and around the dogs' necks. Now Sgt. Lowe was worried about his companion. He tried hard to reason within himself that there was a way out of this predicament.

The pitbulls were released from the back of the building and sent to help the Spetsnaz, but chased after Moose instead. The dogs were quickly gaining on Moose. Perhaps the greatest negative effect of being reanimated was how much the body slowed down. Moose felt the dogs gaining on her. She could hear the barks getting louder and could now hear them breathing as well. Something didn't feel right to her. She decided her only option was to fight. Moose leaped into the air and turned her body a hundred and eighty degrees. Her forward momentum caused her to slide backwards, leaving short claw marks in the ground. Moose then recognized the odd feeling now. She was hungry. Her energy levels were nearly gone. The need to survive was taking over. The approaching dogs could sense an unfamiliar presence. They stopped just short of Moose and began pacing. Moose lowered her head and bore more teeth than ever before. Her growls and barks were haunting, even to the other dogs. Sgt. Lowe could hear Moose and sent a telepathic message of encouragement.

Finally, one of the pitbulls lunged forward towards Moose. The dog leapt at Moose, who had now realized that she could understand what the dog was planning. She could understand its thoughts. The dog was going to catch Moose by the throat and clamp down with his teeth, and vice-like jaws until Moose stopped moving.

Moose, instead of leaping with the dog, lowered her body closer to the ground for a fraction of a second. As the dog passed over her head, Moose raised her head with as many teeth exposed as he possibly could. She tore into the belly of the dog, using the dog's momentum to tear away flesh and fur. A gaping wound was left, and the dog's entrails were falling out before it even hit the ground. It collapsed the second it got all four feet on the ground. Moose was on it in a flash. She tore into the head, needing food more than she ever had since being reanimated. As the need for food lessened, Moose remembered that there were more dogs to fight and Sgt. Lowe needed her help. She looked up from the carcass at the four remaining pit-bulls. They were pacing and barking but not attacking. Moose sent them a message that she was not going to hurt them. The message was not in English, but it was working. The dogs felt safe now and began to calm down. Soon, there was total trust, and the pitbulls were following Moose into battle.

Sgt. Lowe saw this unfold from a distance. He was unsure of what had just happened, but he knew that he must tell Dr. Smitherson about it if he survived.

The dogs charged the Spetsnaz soldiers, who turned and began firing on them. Two of them fall victim to the bullets. They yelped in pain as their bodies rolled to a stop. The wounds were massive, and death came quickly. Moose and the three remaining pitbulls knocked down one of the Spetsnaz and started tearing at his flesh at any point they could get teeth past the body armor. He screamed for help from his comrades, but they could not get a clean shot at the dogs. The bullet would go right through the dog and into the soldier. Moose had discovered that the jacket had a weakness. It could be pushed up with her snout, exposing the soldier's stomach. Moose pinched skin between her front teeth and began pulling backwards. The soldier screamed as his skin stretched to the tearing point. Moose shook her head, and the skin separated from the body.

The soldier pulled off his right Kevlar glove to get his knife, and as soon as he did, a pitbull bit down on his hand and began shaking his arm violently. Bones could be heard snapping, and soon most of the hand was lying on the ground. Blood quickly rushed from the wound, and the soldier's exposed skin turned pale. Shock was setting in as the blood loss mounted and pain became unbearable.

As the soldier's life was leaving him, one of the other Spetsnaz threw off his gloves and took a long knife from a sheath on his belt. He shouted something in Russian and started to move towards the dogs. A single gunshot rang out, and the soldier's head tilted to the left, and blood flowed from his mouth. Sgt. Lowe made it to the AK-47 and shot the charging soldier in the side. A seam in the armor allowed the bullet to pass right through, tearing open the artery that supplies the stomach and nicking his heart. The soldier was alive when he fell to the ground. Life was soon lost, though, as the dogs turned their attention towards him and began biting anywhere they could get through the armor. One of the pitbulls discovered that he could clamp down on the armor and shake violently. The seams would give way under his immense strength, and the armor would fall away. The soldier did not last long after this discovery. Large pieces of him were thrown around by the dogs.

The remaining Spetsnaz ran for cover as they realized Sgt. Lowe was shooting at them. They all dashed behind an out-building and discussed their options. They decided they would walk out in a line and engage Sgt. Lowe. They would rely on training and their marksmanship skills to kill him and their body armor to protect them from the bullets coming at them as he returned fire.

The remaining Spetsnaz ran from cover and started walking towards Sgt. Lowe. A hail of bullets impacted the ground where he had last been sitting, but there was no Sgt. Lowe.

A soldier noticed, he was not there and shouted ceasefire. The soldiers formed a small square with their backs to each other as they slowly moved towards Sgt. Lowes' last position. A single gunshot cracked the air, and one of the soldiers screamed in pain and fell to the ground. He tried desperately to stop the bleeding from the back of his left knee. Sgt. Lowe placed a bullet in the seam of his body armor. The soldier was bleeding to death rapidly. Two of his comrades began pulling off his armor while the third prepared a tourniquet. All three were constantly scanning the area to find Sgt. Lowe's hiding place. Sgt. Lowe thought of shooting again, but knew his position would be compromised. The tourniquet was tightened until the bleeding stopped. The wounded soldier secured his weapon and lay back on the ground as the other three once again started moving, now forming a triangle with their backs facing each other.

The dogs had taken notice of the wounded soldier and were now charging him at a quick trot. The soldier rolled onto his stomach and assumed a prone firing position. His training took over, and he ignored the pain and weakness from blood loss as he aligned his sights on the lead dog. He shot at the dog, and it yelped as it tumbled for several feet and came to a stop on its side. Its eyes were still staring at its target, but life was no longer in them. The soldier drew a bead on the next closest dog and quickly fired. The bullet hit the dog right between the eyes, causing the top of his head to separate. The dog rolled several times, leaving a trail of blood and brains ten feet long. The remaining pitbull stopped as if to try to help his fallen comrades. He whined as he sniffed the bodies. He ran back and forth between the two and seemed to be compelling them to get up. Moose continued her charge.

She was identified now as the greatest threat, so the soldier turned his sights towards her. A shot rang out, and Moose fell over on her left side. She slid a few feet and stopped with her back facing the soldier. The bullet hit and broke both legs on Moose's left side. The legs were mangled, and Moose could not move them. The soldier carefully aimed his rifle at the back of Moose's head. He took his time, having temporarily forgotten about the last pit bull. Just as the soldier squeezed the trigger, the pitbull passed in front of the sights. He was mortally wounded but could still fight. The dog leapt back to its feet and ran directly at the downed soldier. The soldier emptied the remaining four rounds in his magazine, but they all missed their target. The pitbull was nearly dead when it clamped its jaws together with its teeth sunk firmly into the soldier's neck. The dog died with his jaws locked tight, and the soldier closed his eyes and waited to die of suffocation. It would not take long, and he welcomed the end to his suffering. Moose, having been able to tilt her head back enough to see what happened, lay there. She wanted so much to attack, but her legs completely failed her.

Sgt. Lowe saw his dog fall, and anger replaced logic and reason. He leapt from his hiding position and began firing at the remaining three Spetsnaz. The anger was more than he had ever felt before. He was overcome with hatred for the soldiers as he charged towards them. His magazine emptied, and he depressed the magazine release without taking the rifle from his shoulder. The next magazine was already in his left hand, and he slammed it into the magazine well while nearly simultaneously releasing the bolt. He resumed firing without losing a step. The whole time, enemy bullets were impacting Sgt. Lowe. He could feel pieces of flesh tear away and bones shatter. A round struck him in the face, and he could feel teeth being ripped out. The teeth flew into his line of vision and disappeared.

Sgt. Lowe directed his fire towards the face shields of the enemy soldiers. He knew the shields could not hold up forever. "Finally," he thought to himself as a shield shattered and a soldier fell in a dead heap. His helmet was now filled with the same head, but blood, bone, and brains were on the outside of his skin. The two remaining Spetsnaz spread a little farther apart and dropped to the ground in the prone position to make smaller targets. They continued firing at Sgt. Lowe, who was now so close to them that they began to panic. One stopped firing and looked down to find a hand grenade. The decision was a fatal one as a bullet struck the very top of his helmet. The bullet was stopped by the helmet, but the impact shattered vertebrae in the soldier's neck. A fragment of displaced bone severed his spinal cord, and his brain could no longer tell his heart to beat. He was dead without ever knowing about his injury.

Sgt. Lowe was now just twenty steps from his foe. The Spetsnaz could see his eyes and the determination in him to kill him. He assessed that he couldn't stop Sgt. Lowe with a barrage of bullets, so he aligned his sights with Sgt. Lowe's lower right leg and fires. Sgt. Lowe's tibia shattered as the bullet passed through. The fibula snapped under the sudden strain of supporting all the body weight, and Sgt. Lowe fell toward the ground. He threw his rifle aside and extended his arms and hands to break his fall. Bewilderment set in again as he tried to understand what had just happened. He felt no pain, but suddenly could not stand. He immediately rolled onto his back as he hit the ground, sat up, and tried to stand.

The Spetsnaz soldier watched in a combination of admiration, amazement, and horror as Sgt. Lowe struggled to stand until his dry flesh tore, and bone popped through his pant leg. Sgt. Lowe gave up on standing and began dragging himself on his stomach towards the Spetsnaz. He was growling as he moved just inches at a time, and his right leg was twisted completely around and dragging behind him.

The Spetsnaz quickly assessed his own injuries. He was sore and no doubt bruised from the rounds that impacted his body armor, but he detected no serious injuries. He got up from his position and took the few steps needed to reach Sgt. Lowe. Sgt. Lowe accepted his defeat and lay his head across his now folded arms. The Spetsnaz laughed a little as he aimed the muzzle of his rifle directly at the back of Sgt. Lowe's head and aligned the sight apertures as if he were aiming at a target five hundred meters away. He inserted his right index finger into the trigger guard and reaffirmed his sight picture when he heard from directly behind him, "We got this one, Sarge." He turned to see three RAS staring at him. They were burned in places and smelled of burnt metal and electrical components. All of them had singed hair and broken bones. The soldier realized too late that the RAS was quickly advancing on him. His rifle was ripped from his hand before he could fire a shot. Pain coursed through his body as the RAS pulled and twisted, trying to expose flesh. His bones snapped and joints dislocated before the armor could be pulled away. He was in so much pain that he reached with his last bit of strength and unfastened the straps on his helmet and pushed it off. There was only a second more pain as teeth tore into his neck and head.

Sgt. Lowe could now hear tanks and trucks approaching. He heard orders being barked out in English and knew the plan worked. Kasmidod's lines were broken, and soon his palace would fall. Tanks took up positions surrounding the building as members of the 101st circled in front of the tanks and slowly moved in. The lack of cover made everyone tense, but no one failed to do their job. As crews prepared to assault the building, recovery teams began collecting the RAS.

Some had to be captured with bite poles, while others allowed themselves to be bound and gagged and even climbed into the trucks. Weapons systems and pieces of RAS were collected as well. Nothing related to the RAS that could be found was left except Sgt. Lowe and Specialist Hunter, who had by this time found Sgt. Lowe, and helped him stand, using a sniper rifle as a crutch. They both made their way to Moose, who wagged her tail and whined with excitement as they spoke to her telepathically and Sgt. Lowe patted her on the head several times. There was still work to do, but Sgt. Lowe knew he could not be part of it. Hunter helped Sgt. Lowe put Moose on his shoulders, and they slowly walked towards the column of trucks and climbed on one. Specialist Hunter laid her head on Sgt. Lowe's shoulder while they sat on the bench seat, waiting for the long trip home.

As the RAS was being removed from the battlefield, living soldiers were wondering why there was no enemy fire coming from inside the building. An hour passed, and still there was nothing. Company commanders met and decided that troops would have to enter the building, but there was a belief that it was a trap. A message requesting volunteers to make it around to the hundreds of soldiers and within minutes, a group of soldiers was assembled and readied to make entry into the building.

The teams were divided into six twenty-man teams. Each team was assigned a leader and a door to breach. Sergeant First Class Stephanie Collins was selected to lead the team entering through the door opened by Sgt. Lowe. She had been an Infantry Platoon Leader for three years and had been in combat for almost the entire time. She was good at motivating men and women to fight in the most difficult circumstances. She called the other nineteen members of her squad around her and told them to expect to be ambushed as soon as they entered the opening. Her soldiers conjured up visions of gray walls and machine guns firing at them from previously constructed fighting positions. They all knew most were going to die, and yet each one believed he or she would not be one of those killed.

The breach created by Sgt. Lowe was big enough for two soldiers to pass through at once, so the team was divided and ten lined up along each side of the opening. Sfc. Collins counted down from three, and right after she said one, the squad quickly moved inside. Explosions could be heard as other teams used explosives to open their doors, but there was no machine gun fire. In fact, there was no enemy fire at all. The walls were not grey and cold. They were bright and even cheerful. Beautiful works of art hung on the walls and desks were decorated with flowers and pictures of children and spouses. Along the back wall and hanging at eye level were ten carefully framed blankets. The blankets were old, and some were torn and stained, but all were colorful and seemed to scream that they had a story to tell about the person who used and loved them. Various plants were growing from pots, and even a very large aquarium with fish still swimming as though nothing out of the ordinary was happening outside their glass utopia.

Everything in the room, however, was not beautiful. As the shock of the surroundings faded, the smell of death, vomit, and feces overtook the soldiers. Their eyes were drawn to enemy soldiers lying on cots. Some were moaning or weeping, but most were dead.

Next, each cot was a bucket, or a trash can with vomit in it. Feces seeped through the pants of many of the soldiers and pooled on the cots under them. Some have IVs, but the bags of fluids were all empty. Sfc. Collins realized now that there was no danger and began shouting for medics. The medics rushed in expecting to treat wounded American soldiers, and instead were sent to those still alive on the cots. The medics began assessing the soldiers, and one quickly announced that they had the flu. They began caring for the soldiers and carrying them outside for some fresh air. Other squads began reporting that there was no one in the building and there was no enemy contact. Brigade commanders were summoned to a makeshift TOC to discuss where all the leaders who were supposed to be in the building might be.

Sfc. Collins walked to a wall and leaned against it. She needed a moment for her mind to absorb and process what had just happened today. She moved from leaning slightly forward to resting her head with her left ear directly against the wall. Her eyes grew wide as she heard what sounded like footsteps cautiously moving down metal stairs. She could hear whispers but could not make out the words.

Sfc. Collins was listening, and other soldiers began finding evidence that the meeting was cancelled because of the flu outbreak. Lancegoroff himself was gravely ill. There was information that his army was smaller than any could have imagined. There were very few troops left inside the building to guard him, and now most of them were dead or dying from the flu. The only Spetsnaz left was killed outside. It seemed the communists had been stopped, at least temporarily, by unintended germ warfare.

Sfc. Collins was sure now that people were moving behind the wall. She cleared that section of the room and pulled the pin on a grenade. She carefully placed the grenade on the floor against the wall, let the spoon fly, and ran towards the other side of the room. The sound of the grenade exploding was deafening as it echoed repeatedly in the vast room. Sfc. Collins and two other soldiers ran into the hole and realized they were standing on a hidden staircase that extended five floors up and an undetermined number of floors down. Metal floors extended out on each floor. She peered over the railing and spotted Lancegoroff and two men helping him down the stairs. She saw a muzzle flash as one of the men fired a nine-millimeter submachine gun up at her. She stepped back for a second, then leaned over the rail and returned fire with her rifle. One of her bullets struck Lancegoroff in the abdomen, and he fell from the arms of the two trying to help him. One of the two men picked him up and threw him over his shoulder while the other put a near-continuous rate of fire towards the three soldiers above them. The man with Lancegoroff ran down the remaining flights of stairs and vanished out of sight. The other continued to fire until Sfc Collins told one of her men to hit him with the grenade launcher. The soldier leaned forward, pointed his weapon towards the muzzle flash. There was a loud "thud" as he pulled the trigger. The grenade impacted near the man, and he was killed instantly. The explosion moved up the staircase, picking up intensity as it echoed along the way. As the sound moved past them, the three soldiers started running down the stairs.

They had gone ten floors before they found the bottom. Before them, there was a very poorly lit hallway. They could hear a motor in the distance. They ran towards the sound and found that the hall ended in what appeared to be a man-made canal full of water. There was equipment around that would be used to maintain a boat. Sfc. Collins realized Lancegoroff had escaped. She tried to radio the information, but the radio would not function at this depth. The team headed back up the steps.

Lancegoroff and his last man were almost to the ocean. The canal was secretly built for just such an escape. They were traveling in a stolen Navy stealth speedboat. It was reported as sunk but had been ready for months if Lancegoroff should need it. The boat sped towards Cuba, where there was a secret secondary headquarters for the communists. Larzarie, the last man, radioed ahead, and the medical team was waiting to greet them. The boat was fast, but it still took hours to reach Cuba. Lancegoroff was all but dead now. The bullet wound combined with the flu would be his undoing. People were lifting him from the boat, but he ordered them to put him down. He reached into his pocket and pulled out a syringe containing a dose of the stolen Ichor. With nothing to lose, he removed the cap and pushed the needle into his arm. His thumb pressed down on the plunger, and the serum entered his bloodstream. It killed him as soon as he finished the injection. Though dead, activity still took place inside him, just like with all the RAS that had been reanimated. Something different happened inside Lancegoroff, though. The Ichor found the flu virus and began to merge with it. Cells from each slammed into each other and became one larger cell. This happened repeatedly. Less than a minute after death, Lancegoroff opened his eyes. He moved them back and forth a few times, then opened his mouth to smile. A thick black liquid oozed from between his teeth as he got to his feet. Memory came back to him, and he realized he had been successful. He needed to get energy now. Reanimation was draining. He grabbed one of the medical workers and tore out a piece of flesh from her neck.

He was shocked by the taste and let her fall to the ground. She grabbed the wound but couldn't stop bleeding. Lancegoroff turned away to better assess what had happened. As he was looking away, the woman bled to death. Lancegoroff turned around, knowing he must finish what he started. He was startled to see the woman standing back up. She looked confused, and she had no color, but she was moving. He realized what had happened. She had been reanimated. He smiled as he realized his newfound power. He could reanimate with just a bite.

Doctor Smitherson's Notes

As the RAS began to return to his laboratory, Dr. Smitherson decided he needed to take a few minutes to review his notes and write down some questions he needed to answer or have answered for him. He took some time to organize them and decided to give each member of his team a copy. This was difficult because he had never shared his notes before.

Why is energy needed as soon as a corpse is reanimated?

Why did his brother try to eat him right away?

Can energy be provided during the reanimation process?

Why does the energy from human flesh only work on humans?

Why does the energy from the flesh of an animal work only on the same species?

Why does Ichor seem to be able to reanimate any species?

Can these questions be scientifically tested and answered ethically?

Why do some reanimated specimens have much better memory and control than others?

How can this be tested?

If the assumption that a reanimated corpse is a machine or property is correct, could he transplant the brain stem of his brother into a corpse? Would it be his brother or just a machine?

What will be the process for maintaining a division of reanimated soldiers during times of peace?

Would it be ethical to buy the remains of deceased humans to feed them?

Should they be destroyed?

Can they be frozen until needed again?

Can Sgt. Lowe be destroyed? Is he a human with different nutritional needs? Is he a machine, property of the United States Government? Is he a monster?

Can the parts of a destroyed RAS be used to repair one that is still functional but in need of repair? Is that ethical since the donor never agreed to be used for spare parts?

To my team:

I have struggled with many of these questions since I started my work on reanimation. My original intent was to answer questions as work progressed and to not move on until the question was answered. Unfortunately, the immediate military needed to bypass questions that now must be answered.

Just a few short weeks ago, I would have had no objection to destroying every reanimated being we have created. Now, after Sgt. Lowe and Specialist Hunter, I am struggling with destroying any of them. How do we decide on which ones to keep and which ones to destroy? Are there more like Sgt. Lowe out there?

My heart is heavy with the weight of the questions we must answer. This is a feeling that I have never experienced before. I have always looked forward to finding answers. This process has made me question everything I believed professionally, ethically, and spiritually.

We have a great deal of work to do, so I am breaking us up into groups. Sam in Ben will screen the audio and video from the battle. There are captured videos from security cameras placed around Lancegoroff's headquarters. Those will need to be screened as well. We know how the RAS kills, so that is not important. I want you to look for three specific items. A film that will help us understand how hard it is for the enemy to kill the RAS. Look for signs of any RAS that exhibit signs of memory or that appear to know what they are doing. And last, look for evidence that some of the RAS were able to properly interact with troops. Make a copy of whatever you find. I want to view the video before we decide which RAS is to be destroyed for study.

The rest of us will start making repairs to wounded RAS that do not need limbs transplanted. I do not want to do that until we answer the ethical questions. We will put the repaired

RAS in cold storage until we decide how we are going to maintain them. Once the repairs are done, we will start experimenting with food sources and making animals work for RAS.

We will also start answering some of my questions. I want Doctors Cavert and Chi to start working on ethics myself and Dr. Colman will start experiments. I am going to interview Sgt. Lowe tonight, and we will meet in the lab at 5:00 AM

Dr. Smitherson interviewed Sgt Lowe in a one-on-one setting. He not only recorded the interview but also took meticulous notes. The back and forth of questions and answers was so intense that even the dead participant looked exhausted.

After the interview, Sgt. Lowe was taken to the repair and storage building. He and Specialist Hunter were the last RAS troops to be repaired. Sgt. Lowe looked at an open door with what appeared to be a mist coming from it. Through the mist, he could see his fellow RAS members standing in formation. They were motionless and covered in thick plastic. Sgt. Lowe and Specialist Hunter were covered with plastic. The feel of the large, thick bag being pulled over him made Sgt. Lowe feels like the end was near again.

Both were moved into their assigned places in the freezer. As Dr. Smitherson was about to walk away, he heard plastic tearing. He turned around and saw that Sgt. Lowe and Specialist Hunter just tore holes in their bags and extended a hand through the holes. Sgt. Lowe's left hand was now extended through his bag and holding tightly to Specialist Hunter's right hand. She had extended it through a hole in her bag. They were staring at each other through the plastic.

Dr. Smitherson picked up one of the plastic bags from a pile of them and unrolled it on a table. He then picked up a scalpel and cut off about two feet from the bottom of the bag, went into the freezer, and carefully wrapped the two hands together. Next, Dr. Smitherson called for Moose to be brought in. Her legs were repaired with metal screens and duct tape. Moose did not want to enter the freezer at first, but Sgt. Lowe called her in and told her to lie down. Moose lay down across the feet of Sgt. Lowe rested her head on the boots of Specialist Hunter. Dr. Smiterson and a guard used the plastic bag left from the wrap and tucked it around Moose. The guard patted Moose on the head and said, "Good dog." Upon exiting the freezer, Dr. Smitherson told the guard to close and seal the door. Dr. Smitherson returned to his office and recorded the incident of the freezer in his notes. He decided to sleep on the sofa in his office. He was more tired than he could remember ever being in his life until now. He set an alarm for 4:00 AM. Five hours of sleep, he muttered aloud. That will be nice. He turned the light out and made himself comfortable on the sofa.

"Good night, Barney," he said without words.

"Good night, brother," came the soundless reply.

Made in the USA
Monee, IL
06 November 2026

68117942R00125